International Conference on Green and Renewable Energy Resources (ICGRER 2016)

IOP Conference Series: Earth and Environmental Science
Volume 75

Solo, Indonesia
14 – 15 November 2016

ISBN: 978-1-5108-4715-6
ISSN: 1755-1307

Printed from e-media with permission by:

Curran Associates, Inc.
57 Morehouse Lane
Red Hook, NY 12571

Some format issues inherent in the e-media version may also appear in this print version.

Printed by Curran Associates, Inc. (2017)

For permission requests, please contact the Institute of Physics
at the address below.

Institute of Physics
Dirac House, Temple Back
Bristol BS1 6BE UK

Phone: 44 1 17 929 7481
Fax: 44 1 17 920 0979

techtracking@iop.org

Additional copies of this publication are available from:

Curran Associates, Inc.
57 Morehouse Lane
Red Hook, NY 12571 USA
Phone: 845-758-0400
Fax: 845-758-2633
Email: curran@proceedings.com
Web: www.proceedings.com

Table of contents

Volume 75

International Conference on Green and Renewable Energy Resources (ICGRER 2016)

14–15 November 2016, Solo, Indonesia

Accepted papers received: 5 July 2017
Published online: 9 August 2017

Preface

Solar Energy/Solar Cell

Biomass & Bioenergy/Ocean and Tidal Energy

Miscellaneous Product

Preface

The International Conference on Green and Renewable Energy (ICGRER 2016) was held November 14-15th, 2016 at Solo Paragon Hotel, Surakarta, Central Java, Indonesia. This conference is organized by Faculty of Mathematics & Natural Science, Sebelas Maret University.

The scope of parallel classes are solar energy/solar cells, biomass and bio energy, and miscellaneous product. There are 60 presenters were presented and discussed in plenary and parallel session, therefore, a total 28 manuscripts have been peer reviewed with good quality.

We are pleased to have Prof. Dr. Sri Adiningsih, M.Sc., Ph.D (Chairman of the Presidential Advisory Council) as our outstanding keynote speaker. We also glad ti have our invited speakers Prof. Renate Egan (University of New South Wales Australian PV Institute, Solinno Pty, Ltd), Dr. Adhi Wibowo (Direktur PPTEK R&D EBTKE Kementerian ESDM), Ir. Hj. Sripeni Inten Chayani, M (President Director PT. Indonesia Power), Prof. Dr. Ir. Kamaruddin Abdullah (Universitas Darma Persada), and William Papet (Business Development at MPS Consortium).

We express our gratitude to participants from all researchers and students for join this conference. We also appreciate for sponsors, reviewers, and all member of the organizing committee which prepare this conference. Our gratitude also goes to the IOP Publisher for cooperation in prepare this proceeding. We do realize that the success of the conference depends on both participants and who have worked with us.

Prof. Ari Handono Ramelan

Chairman of ICGRER 2016

Dean of Faculty Mathematics & Natural Sciences

IOP Conf. Series: Earth and Environmental Science **75** (2017) 011002 doi:10.1088/1755-1315/75/1/011002

Peer review statement

All papers published in this volume of *IOP Conference Series: Earth and Environmental Science* have been peer reviewed through processes administered by the proceedings Editors. Reviews were conducted by expert referees to the professional and scientific standards expected of a proceedings journal published by IOP Publishing.

IOP Conf. Series: Earth and Environmental Science **75** (2017) 012001 doi:10.1088/1755-1315/75/1/012001

Study of Performance of *Coaxial Vacuum Tube Solar Collector* on Ethanol Distillation Process

Sutomo[1], A H Ramelan[2], Mustafa[3], and T Tristono[4]
[1,3] Mechanical Engineering Department, Universitas Merdeka Madiun, Madiun, 63133, Indonesia
[2] Faculty of Mathematics and Natural Science, Sebelas Maret University, Surakarta, 57126, Indonesia
[4] Informatics Management Department, Universitas Merdeka Madiun, Madiun, 63133, Indonesia

E-mail: omotkus26@gmail.com

Abstract. Coaxial vacuum tube solar collectors can generate heat up to 80°C is possibly used for ethanol distillation process that required temperature 79°C only. This study reviews the performance of coaxial collector vacuum tube used for ethanol distillation process. This experimental research was conducted in a closed space using a halogen lamp as a solar radiation simulator. We had done on three different of the radiation values, i.e. 998 W/m^2, 878 W/m^2 and 782 W/m^2. The pressure levels of vacuum tube collector cavity in the research were 1; 0.5; 0.31; 0.179; and 0.043 atmospheres. The Research upgraded the 30% of ethanol to produce the concentration of 77% after distillation. The result shows that the performance of coaxial collector vacuum tube used for ethanol distillation process has the negative correlation to the level of the collector tube cavity pressure. The productivity will increase while the collector tube cavity pressure decreased. Therefore, the collector efficiency has the negative correlation also to the level of collector tube cavity pressure. The best performance achieved when it operated at a pressure of 0.043 atmosphere with radiation intensity 878 W / m^2, and the value of efficiency is 57.8%.

1. Introduction

Human life on the earth always needs energy. It has an implication that human population growth will affect the amount of energy required. According to the International Energy Agency-IEA, world energy demand will increase steadily. It predicted, the average up to 2030 is 1.6% per year. The whole nation in the world will continue facing energy supply problems. In another hand, especially the dwindling amount of fossil energy, while eighty percent of the energy supply in the world using fossil fuels. We know that burning fossil fuels release harmful emissions, such as carbon dioxide, nitrogen oxides, aerosols, etc. that can pollute the environment. We need to have a framework to develop energy sources that do not pollute the environment. The alternative solution is to use environmentally friendly renewable energy as an energy source.

The most abundant of the renewable energy sources is the solar radiation, especially in the area around the equator. Solar radiation is the main source of the energy system that can affect the Earth's natural cycles as well as support all life on the earth [1]. The technology applied to use the energy of solar radiation is classified into two types. It was namely photovoltaic technology and solar thermal technology [2].

Ethanol as a renewable energy source is very potential to be also developed because it can be produced from carbohydrates contained in the plant, (cassava, maize, sorghum, etc.) or molasses (sugar production waste). Ethanol can improve combustion efficiency, especially in motor Otto. It because of ethanol contains approximately 35% oxygen. It has a higher octane number (112) also if it's compared to gasoline that has the octane value less than 90 only [3]. The low-level ethanol (10%-20%) can be produced in several ways, one of which is a biological process of fermentation. Increasing the level of ethanol can be done by separating the water contained therein. Distillation is the process of components separation based on their boiling points. Pure ethanol boiling point is 79°C and water boiling point at a standard condition is100°C [4].

Therefore, this research is vital because the aim of this study regarding the performance of coaxial vacuum tube collector used for ethanol distillation process. None of the researchers applied the coaxial vacuum tube collector for ethanol distillation process that required a temperature of 79°C only. As we have already known that the vacuum tube collectors generate heat up to 80°C. This distillation technique based on the differences in boiling point or melting point is very useful to upgrade the 30% of ethanol to the concentration of 77% [5].

The previous researches related to this study as the comparison of ours are [6-14]. They had talked about solar collector or water heater efficiency.

This paper organizational structure is as the following: Chapter 1 talks about the introduction, chapter 2 for analysis and efficiency of evacuated tube collectors, methods in chapter 3, result and discussion in chapter 4, and chapter 5 for the conclusion.

2. Experimental

2.1. Analysis and efficiency Evacuated Tube Collectors (ETC)

The Solar thermal collector is a heat exchanger system that utilizes radiation as a primary energy source. It collects the heat and transfers to the circulating fluid in it (fluid can be air, water, or oil). Basically, the solar thermal collectors are classified into two types i.e. non-concentrating or stationary and concentrating. The stationary solar thermal collector system was classified into three categories, i.e., 1. Flat-Plate Collectors (FPC), this type uses the absorber, 2. Compound Parabolic Collectors (CPC). It uses a concentrator component that made of the high transmissivity material. This concentrator capable is focusing the energy of solar radiation on a receiver. 3. Evacuated Tube Collectors (ETC) [17].

Principally, the Evacuated Tube Collectors (ETC) put the absorber in the vacuum glass tube. The advantages of this collector due to the high heat transfer efficiency. It because the heat loss factor is relatively low. It due to the air trapped between the absorber and the glass tube conditioned in a vacuum space.

The Evacuated Tube Collectors can produce fluid temperature 50°C-200°C. We improve the effectiveness of the absorption of radiation absorber using the metal having high conductivity (copper) that painted black.

There are several types of absorber employed in the evacuated tube collectors, one of which is Co-axial Evacuated Tube Collectors. This collector has the advantage absorber cylindrical shape. The total amount of radiation of incident ray which is perpendicular to the absorber unchanged. It has implications on the total effective area of heat absorption is always equal.

At steady state, the entire effective heat rate at the collector is equal to the rate of heat absorbed by the heat absorber plate minus the heat loss directly or indirectly discharged into the environment. According to [18], the Equation (1) defines the total effective heat rate of a collector.

$$Q_u = A_c[\varsigma - q] = \dot{m} c_p [T_o - T_i] \quad (1)$$

A_c = area collectors, T_i = the fluid temperature of the inlet, T_o = the fluid temperature of the outlet, = the mass flow rate of the fluid, and c_p = specific heat of the fluid.

The Equation (2) presents the efficiency of the collector.

$$\eta = \frac{Q_u}{A_c I_i} = \frac{Absorbed\ solar\ energy}{Incident\ solar\ energy} - \frac{Energy\ losses}{Incident\ solar\ energy} \quad (2)$$

Q_u = total effective heat rate, A_c = absorber plate area, I_i = the intensity of radiation absorbed by the absorber plate, η the efficiency of solar collectors. [19].

Equation (1) substituted into Equation (2), the collector efficiency becomes the Equation (3).

$$\eta = \frac{\dot{m} C_p (T_0 - T_i)}{A_c I_i} \quad (3)$$

2.2. The distillation process

This experimental research was held in closed space using a halogen lamp as a solar radiation simulator. In this test, we used a coaxial vacuum tube solar collector consisting of transparent glass tubes with a diameter of 48 mm, the length 750 mm, and tube heat absorber made of copper with a diameter of 30 mm, the length of 710 mm. The tube heat absorber was painted black.

The definition of the vacuum is a space with a gaseous pressure than 1 atmospheric pressure. The meaning of the absolute vacuum is exactly zero gaseous pressure, but this is as a theoretical concept only. We simply call "vacuum" to use the term of the partial vacuum. The value of a partial vacuum refers to the proximity degree of an absolute vacuum [11], [12]. We created a vacuum using a pump to remove the gas from the tube. The implication was the gaseous pressure be reduced approaching the exactly zero gradually. We researched 4 different degrees of vacuum tube value that had the gaseous pressure less than 1 atmosphere.

During the experiment, we kept the level of 30% ethanol in the reservoir vessel in a permanent position. This was done ensure the inlet pressure of vacuum tube solar collector was always constant. Of course, this had implications for the precision outcome of the outlet flow volume. It can control the temperature not higher than 80°C also.

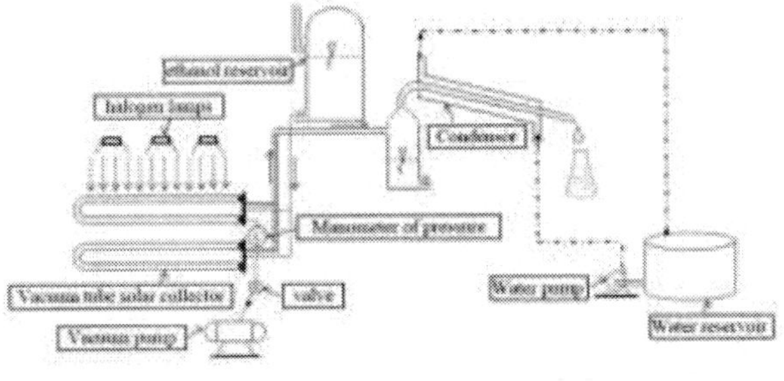

Figure 1. The distillation scheme using coaxial vacuum tube solar collector [5].

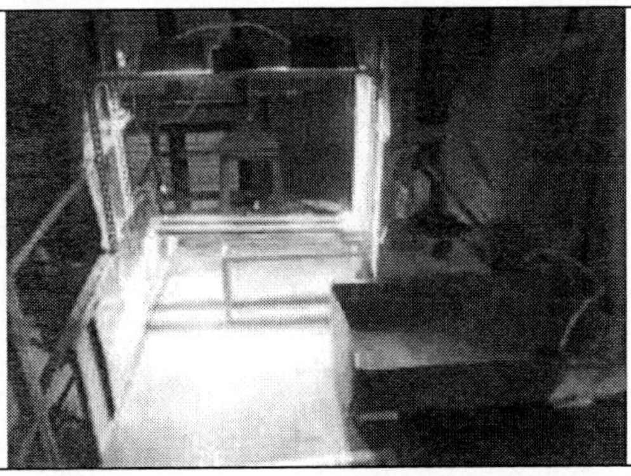

Figure 2. The distillation using coaxial vacuum tube solar collector

At the beginning step of the experiment was done at the pressure of tube collector 1 atmosphere and radiation intensity 998 W/m^2. The activated was done as data acquisition for one minute, and the real testing was for 90 minutes. The volume of distilled ethanol produced was measured after the absorber temperature reaches 35°C. We repeated from the beginning step for radiation intensity 878 W/m^2, and 782 W/m^2. The next step was repeated for all of the steps from the beginning using the pressure of tube collector 0.5, 0.315, 0.179, and 0.043 atmospheres.

The primary data obtained from the test include radiation intensity, pressure tube collectors, inlet volume flow of ethanol to the collector, glass tube temperature, absorber temperature, the inlet and outlet of ethanol temperature of the collector, environmental temperature, and volume ethanol distillation. We collected temperature data using sensor LM 35 and ARDUINO, roared model MECA 2560 R3. We have programmed the setting of the microcontroller for high precision data.

3. Result and Discussion

The Fig. 3 shows that evaporation temperature rise in ethanol outlet small relatively when associated with each level of the cavity pressure of tube collector declines. It occurred in all of the radiation intensity of the experiments. For increased radiation intensity and a decrease in the level of cavity pressure of the collector tube, distillation process affects the amount of ethanol produced. The total energy absorbed by the collector, on each of the radiation intensity 782 W/m^2, 878W/m^2, and 998W/m^2 are 21.25 Joule/second (J/s), 23.86 J/s, and 27.12 J/s, respectively

.

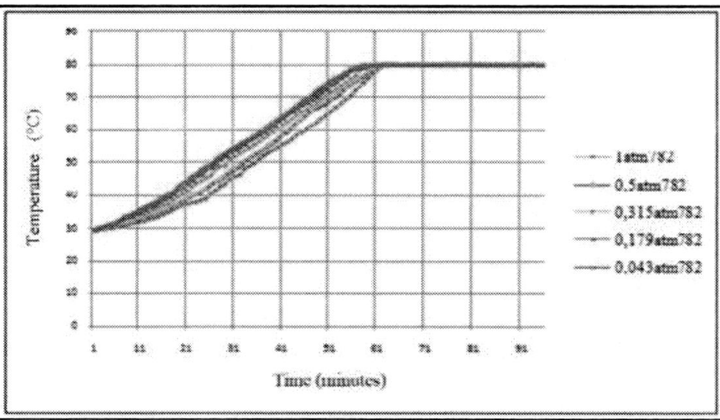

Figure 3. The overview of the evaporation temperature of the radiation intensity 782 W/m^2

Table 1. The total effective heat rate (Joule/second) at a collector and the efficiency

Pressure of tube collector (Atmosphere)	Radiation intensity					
	782		878		998	
	(W/m^2)	%	(W/m^2)	%	(W/m^2)	%
1	13.839	52.89	15.417	52,48	16.666	49.91
0.5	14.013	53.56	15.596	53,09	16.919	50.67
0.315	14.342	54.81	15.934	54,24	17.255	51.67
0.179	14.376	54.94	16.121	54,88	17.621	52.77
0.043	14.988	57.28	16.987	57,02	18.856	56.47

Fig. 4. presents the total of energy collected to achieve the ethanol evaporation. The lower of the tube collectors cavity pressure has implications evaporation temperature is reached faster.

IOP Conf. Series: Earth and Environmental Science **75** (2017) 012001 doi:10.1088/1755-1315/75/1/012001

Fig. 5. shows that the cavity pressure inside the tube collector has a negative correlation with the rate of heat energy utilized in the distillation process. Reviewing of these conditions, we can conclude that the lower cavity pressure of the tube collector has implications to the smaller heat loss.

Total heat utilized in the process of ethanol distillation affected by the level of collector tube cavity pressure. The lower pressure in the cavity tube collector has implications to the amount of heat used for distillation process increases. This condition happens in all experiments of the radiation intensity i.e. 782 W/m^2, 878 W/m^2 dan 998 W/m^2. The research results also present that ethanol productivity affected by the level of tube collectors' cavity pressure. Ethanol productivity increases if the value of the pressure tube collector is lowered. The lower level of the tube collectors cavity pressure increasing the efficiency. In other words, if we distillate ethanol by vacuum collector tube cavity, better we use low-pressure level of collector tube cavity. If carrying out this way, we can save energy, improve productivity and increase efficiency.

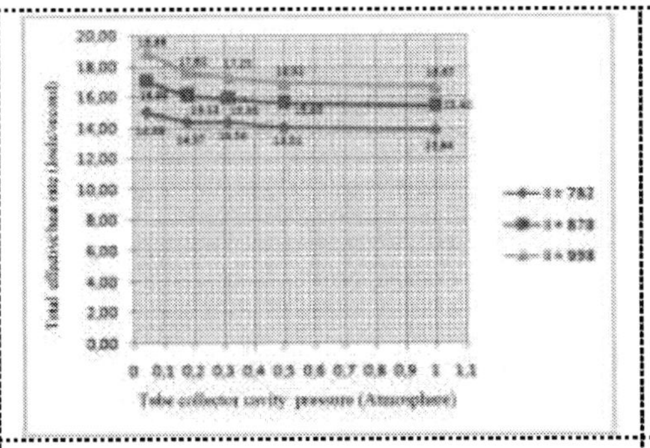

Figure 4. Total of energy collected to achieve the ethanol evaporation

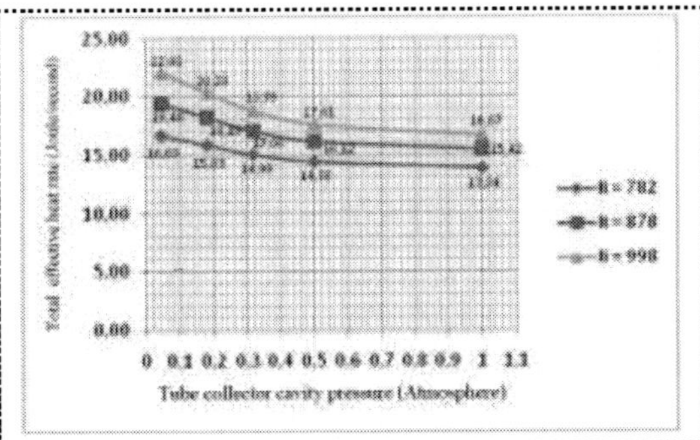

Figure 5. The total of energy utilized in the distillation process

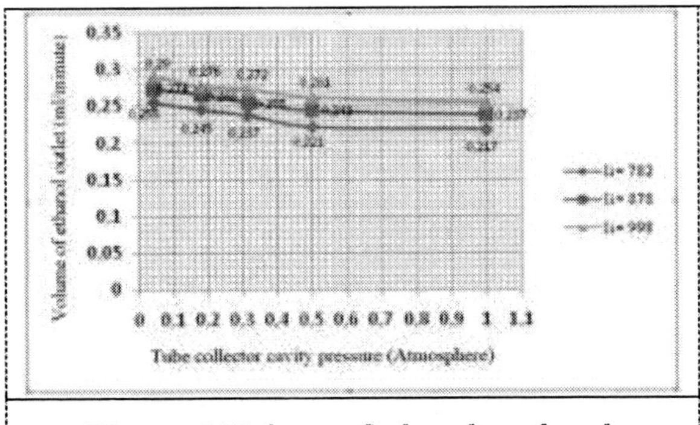

Figure 6. Volume of ethanol produced.

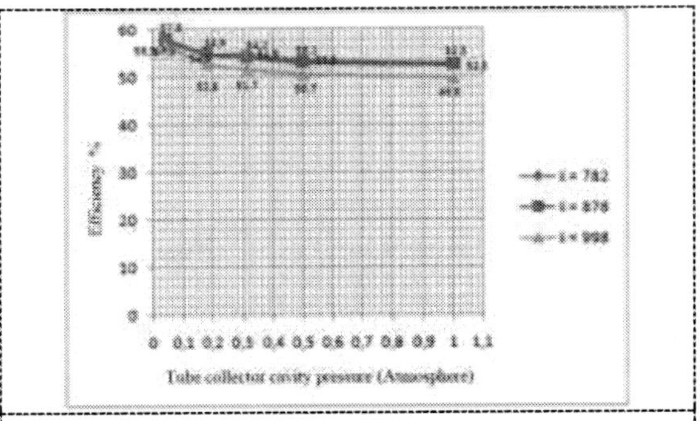

Figure 7. The efficiency of the tube collector

A vacuum cavity of the collector tube contributes as an insulator. The smaller vacuum level impacts to the greater thermal resistance values. The lowest number of the degree of vacuum cavity provides the best insulation for the propagation of heat out. The great amount of the heat more focused for boiling the ethanol.

4. Conclusion

Based on the result of this research, we conclude that the performance of coaxial vacuum tube collector used in this ethanol distillation process was influenced by the level of tube collectors' cavity pressure. The lower of the pressure level of the collector tube collectors have implications for the better performance. The best performance achieved while it operated at a pressure rate of 0.043 Atmosphere with radiation intensity 878 W / m^2 and the efficiency value is 57.8%.

Acknowledgments
This research was performed at Biofuel Laboratory of Sebelas Maret University, Surakarta– Central Java, Indonesia. We are grateful for the generosity of Prudential to Ibu Sayekti Wahyuningsih. We wish to thank R. Setiahadi for the financial support, and to my family also.

References

[1] Ricci M, Bocci E, Michelangeli E, Micangeli A, Villarini M, and Naso V 2015 *Energy Procedia* **82** 744–751
[2] Putra A T I D G 2012 *Logic* **12** No 1
[3] Gusmarwani S R 2010 *Jurnal Teknik Kimia Indonesia* **9** No 3 77-82
[4] Primy R N 2012 *Sain dan Seni ITS* **1** No 1
[5] Khopkar S M 2003 *Basic Concepts of Analytical Chemistry* (Jakarta: UI Press)
[6] Shah L J and Farbo S 2004 *Appl. Energy* **78** 371-395
[7] Kim Y and Seo T 2007 *Renewable Energy* **32** 772-795
[8] Morrison G L 2009 *Sol. Energy* **83** 49-56
[9] Budihardjo I, Morison G L, and Behnia M 2007 *Sol. Energy* **81** 1460-1472
[10] Budihardjo I and Morrison G L 2009 *Sol. Energy* **83** 49-56
[11] Glembin J, Rockendorf G, and Scheuren J 2010 *Sol. Energy* **84** 1137-1146
[12] Mulyono 2010 *Jurnal Teknik Industri* **11** No 1
[13] Zhang X, You S, Ge H, Gao Y, Xu W, Wang M, He T, and Zheng X 2014 *Energy Convers. Manage.* **84** 80–87
[14] Ayala A A, Rodríguez G M, Núñez M P, Ramírez A R U, and Muñoz A G 2015 *Energy Convers. Manage.* **94** 472–481
[15] Duffie J A Beckman W A 1980 *Solar Engineering of Thermal Processes* (New York: Willey-Intrscience)
[16] Finney T Ross L 1997 *Calculus and Analytic Geometry* ed 6th (Jakarta: Erlangga)
[17] Winter Ed 1990 *Solar Collectors Energy and Materials* (Cambridge: MIT Press)
[18] Hayek M, Assaf J, and Latief W 2011 *Energy Procedia* **6** 618-626
[19] Liang R, Ma L, Zhang J, and Zhao D 2011 *Sol. Energy* **85** 1735–1744

IOP Conf. Series: Earth and Environmental Science **75** (2017) 012002 doi:10.1088/1755-1315/75/1/012002

A Study on Structure/Phase Transformation of TiO_2 nanorods at Various Annealing Temperatures

S Wahyuningsih[1], A H Ramelan[2], and L N M Z Saputri[1]
[1] Inorganic Material Research Group, Faculty of Mathematics and Natural Sciences, Sebelas Maret University, Ir. Sutami street 36A Kentingan Surakarta Indonesia
[2] Electronic Materials and Energy, Faculty of Mathematics and Natural Sciences, Sebelas Maret University, , Ir. Sutami street 36A Kentingan Surakarta Indonesia

E-mail: sayekti@mipa.uns.ac.id

Abstract. Synthesis of TiO_2 nanorods was conducted through mechanochemical of ball milling at speed of 175 rpm for 5 hours and strong base 12 M NaOH reaction by hydrothermal at 150°C overnight on variation annealing at 400, 500, and 600°C. Material characterization were performed by X-Ray Diffraction (XRD), Transmission Electron Microscoe (SEM), and Surface Area Analyzer (SAA). Strong base reaction by hydrothermal showed the presence of anatase, brookite, and rutile phase for annealing up to 400°C. The diffraction pattern of annealing at 500 and 600°C contain peaks of both brookite and rutile phase. Morphology transformation of TiO_2 to form nanorods TiO_2 was showed by rod-shape from TEM micrographs and increase surface area into 111.9 m^2/g.

1. Introduction

Over the past decade, one dimentional (1D) nanostructured of metal oxides have attracted much attention due to the advantage of chemical, structural, and physical properties that can be obtained [1]. In particular, TiO_2 nano-materials such as nanorods are widely used in solar cells and photocatalytics applications [2-3]. TiO_2 is well known has three polymorphs including rutile, anatase, and brookite. Anatase and brookite are metastable and can transform to rutile when they were heated, while rutile is the stable one. Many methodes have been developed for the successful synthesis of TiO_2 nanorods such as sol gel, template, sputtering, electrochemical, and hydrothermal. Recently, nanorods derived from the alkali treatment of TiO_2 nanoparticles by hydrothermal method have attracted interest due to the beneficial method that can be obtained in large surface area and good crystallinity [4].

Effect of thermal treatment in TiO_2 and TiO_2 nanorods synthesis gives ways to control grain size, morphology, phase composition via experimental parameters such as time and temperatures. According Rath et al. (2009) [5], there are three competition that occur in phase transformation of anatase to rutile titania structure including grain growth of anatase phase, conversion of anatase to rutile, and grain growth of rutile phase, while Hu et al. (2003) [6] showed that phase transformation to rutile can occur from nanosize anatase accompanied brookite phase. This phenomena still has little information in literatures. Therefore, the effect of annealing temperature on the phase tranformation with nanorods morphology were investigated in this work.

2. Experimental

In Anatase-Brookite (A-B) phase nanocrystalline TiO_2 was used as precursor to obtain TiO_2 nanorods by milling, mechanochemical process. Milling was conducted in ball mill for 5 hours with a ball-to-powder weight ratio of 20:1 at speed of 175 rpm. 1.5 g of as-milled powders were added to 50 mL of 12 M NaOH and the mixture was stirred for a hour at ambient temperature. It then was heated by hydrothermal in Teflon-lined autoclave at 150°C overnight. The white precipitate was filtered and washed with 0.1 HCl until pH netral continuosly. The precipitate was dried in an oven at 105°C. Transformed TiO_2 were obtained by annealing of precipitate at 400, 500, 600°C for 2 hours. Phase identification was performed via X-ray diffractometry (XRD) on Bruker D8 Type diffractometer operating at 40kV/35mA using Cu Kα radiation. Particle morphology was observed via electron microscopy (TEM). The volume fraction phase of the samples were calculated from the following equations [7]:

$$W_A = \frac{k_A A_A}{k_A A_A + k_B A_B + A_R} \times 100\% \quad (1)$$

$$W_B = \frac{k_B A_B}{k_A A_A + k_B A_B + A_R} \times 100\% \quad (2)$$

$$W_R = \frac{A_R}{k_A A_A + k_B A_B + A_R} \times 100\% \quad (3)$$

where W_A, W_B, and W_R represent the weight fractions of the anatase, brookite, and rutile phases, respectively; A_A , A_B , A_R are the X-ray integrated intensities of the anatase (1 0 1) peak, brookite (1 2 1) peak, and rutile (1 1 0) peak, respectively; k_A (0.886) and k_B (2.721) are two coefficients. The size of crystallite was calculated by Scherrer equation.

3. Result and discussion

TiO_2 nanoparticles by ball milling process that have been formed are used as a precursor in the synthesis of TiO_2 nanorods. Synthesis of TiO_2 nanorods draws on research Wahyuningsih et al. (2016) [8] by hydrothermal process with development of various annealing temperatures. The various concentrations of alkali NaOH also affect the structure of TiO_2 nanorods. In the hydrothermal process with a concentration of 8 M NaOH was not obtained rod structure yet, while at concentrations of 10 and 12 M NaOH obtained structural rods (rods) and the formation of rod structure is more dominant on the concentration of NaOH 12 M [8]. According to Liu et al. (2009) [9], hydrothermal process includes synthesis of TiO_2 in alkali solution, turnover proton alkali ion in solution, dried, and washed with acid.

TiO_2 undergo hydrolysis when reacted with NaOH 12 M lead to termination of the polymer chain TiO_2 (depolymerization). OH is highly reactive of NaOH that attacks bonding Ti-O-Ti. OH replace the damaged O-Ti Ti-OH bonding while loose O-Ti will bind Na^+. Neutralized with HCl hydrolysis to the polymerization process again with the release of H_2O at the two ends of TiO_2 polymers re-formed. The process of re-establishment of TiO_2 is called condensation polymers [10].

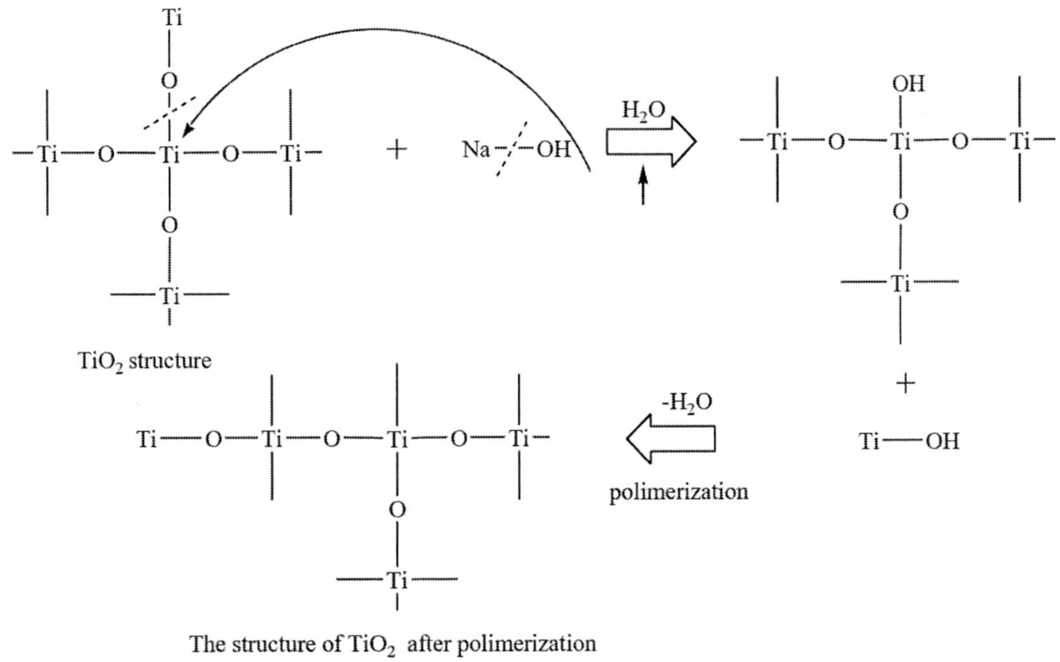

Figure 1. The scheme of mechanism reaction of TiO_2 formation

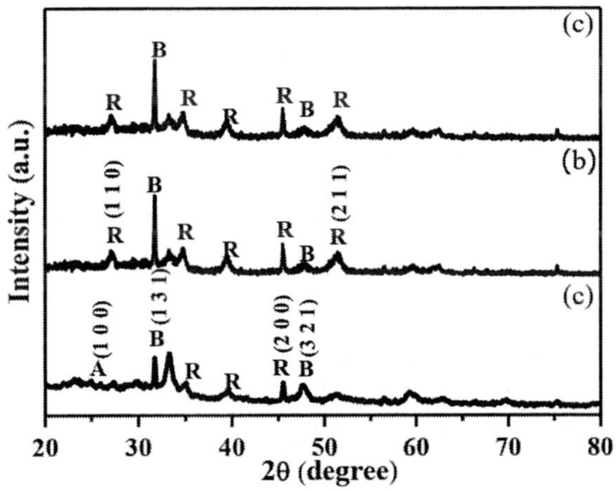

Figure 2. X-Ray diffractogram pattern of TiO_2 by hydrothermal synthesized at annealing temperature of (a) 400, (b) 500, and (c) 600°C. A: anatase, B: brookite, R: rutile.

Figure 2 shows the XRD diffractogram pattern of TiO_2 annealed at various temperatures and there are three phase that obtained including anatase, brookite, and rutile (JCPDS No.76-2486, 76-1934, dan 79-1640). For the sample annealed at 400°C, anatase phase at $2\theta = 25.93°$ (d100); brookite phase at $2\theta = 30.69°$ (d211) and 48.13° (d321); rutile phase at $2\theta = 35.45°$ (d101) which amorphous peak. When the sample annealed at 500 dan 600°C, there is no diffraction peak of anatase phase, but the sample show diffraction peaks of brookite phase at $2\theta = 30.75°$ (d211) and 48.05° (d321); and rutile phase at $2\theta = 27.05°$ (d110), 35.01° (d211), 39.57° (d111), dan 52.11° (d211).

Rhat et al. (2009) [5] explains that the various annealing temperature can affect the structure of TiO_2. TiO_2 thin layer of titanium isopropoxide precursors undergo a phase transformation of anatase into rutile at a temperature of 700-800°C and was obtained pure rutile phase at temperatures of 1000°C. The same results occurred on TiO_2 nanotube in which the influence of the annealing able to bring the rutile phase at a temperature of 600°C and an overall transformation into rutile at a temperature of 800°C [11]. In this study, the rutile phase can be formed at annealing temperatures of 400, 500, and 600°C. Although it has not formed of pure rutile phase, this achievement is of great concern that given in this study of phase transformation into rutile. The rutile of TiO_2 can be achieved if the annealing temperature more than 600°C.

Table 1 shows the crystals size of anatase TiO_2 that annealed at 400°C are about 41.25 nm. Brookite phase TiO_2 increased crystal size is comparable with annealing temperature increases, while rutile phase TiO_2 decreased the crystal size is inversely related to the annealing temperature increases. Table 2 indicates that after hydrothermal and various annealing temperature appear new phase i.e. rutile in which the higher the annealing temperature the anatase phase percentage is getting lower while the rutile phase is increasing. This indicates that the closer the relationship between the influence of the annealing temperature increases to changes in the crystal size and the occurrence of phase transformations.

Table 1. Crystal size of TiO_2 at various annealing temperatures

Annealing temperature (°C)	Phase	B		2θ	Cos θ	D (nm)
		(°)	x 10^{-3}(rad)	(°)		
400	Anatase	0.12	3.44	24.31	0.98	41.25
	Brookite	0.28	4.85	30.40	0.96	29.62
	Rutile	0.19	3.43	35.02	0.95	42.39
500	Brookite	0.22	3.78	30.83	0.96	38.02
	Rutile	0.25	4.59	34.65	0.95	31.66
600	Brookite	0.21	3.71	31.76	0.96	38.88
	Rutile	0.28	4.84	34.50	0.95	29.98

Table 2. Phase percentage of TiO_2 at various annealing temperatures

Annealing temperature (°C)	Phase percentage (%)		
	Anatase	Brookite	Rutile
400	24.48	44.88	30.67
500	-	57.06	42.94
600	-	57.06	42.94

Associated with brookite crystal growth, brookite phase transformation into rutile may occur at the grain boundaries or in the particle brookite. When the core of rutile formed on the inside of the particle brookite, rutile crystals can be increased with increasing annealing temperature. While in this study decreased the crystals size of rutile indicated that rutile core was formed at grain boundaries brookite. An increase in the annealing temperature, brookite crystals grow larger than its size with consequent reduction of the volume fraction of grain boundaries. Because the core of rutile crystals grown at the grain boundaries brookite particles, reducing the volume fraction of grain boundaries at a higher temperature so that it can suppress the growth of rutile crystals. This causes the crystal size of rutile declined against the annealing temperature rise [5,7].

Based on the results of XRD characterization, an increase in the annealing temperature causes phase formed brookite and rutile TiO_2. Crystal size and percentage calculations showed that the phase of maximum yield generated on TiO_2 annealing temperature of 600°C. Subsequently, TiO_2 that annealed at 600°C was characterized by TEM to confirm that it has formed TiO_2 nanorods. TEM imaging results indicated that it has formed rods (rods) TiO2 result of hydrothermal treatment with NaOH 12 M and annealing temperature of 600°C with a particle diameter of 8 nm (Figure 24). The existence of brookite are metastable phase transformation into a stable rutile structure facilitates the formation of nanorods with crystal growth upwards to form rods (rods) [12].

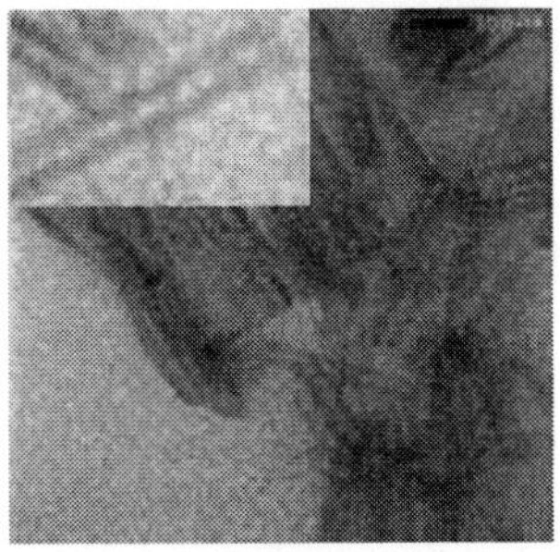

Figure 3. TEM morphology of TiO_2 nanorods by hydrothermal process and annealed at 600°C

Table 3. Surface area of TiO_2 by milling process and TiO_2 nanorods

TiO_2	Methode	Surface Area (m^2/g)
Milling	BET	7.5
Nanorods	BET	111.9

The existence of structural engineering TiO_2 nanorods also increases the surface area up to 14 times. TiO_2 surface area is important to the adsorption of compounds dye (dye). The greater surface area of the adsorption also increased the bounding of dye compound to the TiO_2. The direction of movement of the electrons are regularly when used as a semiconductor. These results are consistent with research conducted by Govindaraj et al. (2014), the structural changes in the form of nanorods were able to improve the characteristics of the surface area of the material [13].

4. Conclusion

TiO_2 nanorods were synthesized by hydrothermal method. The structural/phase transformation of TiO_2 nanorods have been investigated. Strong base reaction by hydrothermal showed the presence of anatase, brookite, and rutile phase for annealing up to 400°C. The diffraction pattern of annealing at 500 and 600°C contain peaks of both brookite and rutile phase. Morphology transformation of TiO_2 to form TiO_2 nanorods was showed by rod-shape from TEM micrographs and increase surface area into 111.9 m2/g.

References

[1] Ji S and Ye C 2008 *Journal of Material Science Technology* **24** 457

[2] Yu H, Bai Y, Zong X, Li X, and Wang L 2013 *Chemical A European Journal* **19** 13569-13574

[3] Paola A D, Cufalo G, Addamo M, Campostrini R, Ischia M, Ceccato R, and Palmisano L 2008 *Colloids Surface* **317** 366-376

[4] Yuli V K, Kovnir, K A, Gavrilov A I, Garshev A V, Frantti J, Labedev O I, Churagulov B R, Tendeloo G V, and Yoshimura M 2006 *Jjournal of Physical Chemistry B* **110** 4030-4038

[5] Rath H, Anad S, Mohapatra M, Dash P, Som T, Singh U P, Mishra N C 2009 *Indian Journal Physics* **83** 559-565

[6] Hu Y, Tsai H L, and Huang C-L 2003 *Journal of the European Ceramic Society* **23** 691-696
[7] Li G, Takamasa I 2004 *Applied Physics* **96** 6663
[8] Wahyuningsih S, Fadhilah G, Hidayat R, and Ramelan A H 2016 *Procedia Chemistry* **19** 632-637
[9] Liu B and Eray S A 2009 *The American Chemical Society* **131** 3985-3990
[10] Ishizaki K, Komarneni S, and Nanko M 1998 *Porous Material: Process technology and applications* (London: Kluwer Academic Publisher)
[11] Yang L, Miao Z, Shiwei S, Jianguo Lv, Xueping S, Gang H, and Zhaoqi S 2014 *Nanoscale Research Letters* **9** 621
[12] Zhu S C, Xie S H, and Liu Z P 2015 *American Chemical Society*
[13] Govindaraj R M, Pandian S, Ramasamy P, and Mukhopadhyay S 2014 *Chemical Technology Reseearch* **6** 5220-5225

IOP Conf. Series: Earth and Environmental Science **75** (2017) 012003 doi:10.1088/1755-1315/75/1/012003

Band-engineering of TiO_2 as a wide-band gap semiconductor using organic chromophore dyes

S Wahyuningsih[1], I Kartini[2], A H Ramelan[3], L N M Z Saputri[1], and H Munawaroh[1]

[1] Inorganic Material Research Group Faculty of Mathematics and Natural Sciences, Sebelas Maret University

[2] Inorganic Chemistry, , Chemistry Department, Faculty of Mathematics and Natural Sciences, Gadjah Mada University

[3] Electronic material and energy research group, Faculty of Mathematics and Natural Sciences, Sebelas Maret University

E-mail: sayekti@mipa.uns.ac.id

Abstract. Bond-engineering as applied to semiconductor materials refers to the manipulation of the energy bands in order to control charge transfer processes in a device. When the device in question is a photoelectrochemical cell, the charges affected by drift become the focus of the study. The ideal band gap of semiconductors for enhancement of photocatalyst activity can be lowered to match with visible light absorption and the location of conduction Band (CB) should be raised to meet the reducing capacity. Otherwise, by the addition of the chromofor organic dyes, the wide-band gab can be influences by interacation resulting between TiO2 surface and the dyes. We have done the impruvisation wide-band gap of TiO2 by the addition of organic chromophore dye, and the addition of transition metal dopand. The TiO2 morphology influence the light absorption as well as the surface modification. The organic chromophore dye was syntesized by formation complexes compound of Co(PAR)(SiPA)(PAR)= 4-(2-piridylazoresorcinol), SiPA = Silyl propil amine). The result showed that the chromophore groups adsorbed onto TiO2 surface can increase the visible light absorption of wide-band gab semiconductor. Initial absorption of a chromophore will affect light penetration into the material surfaces. The use of photonic material as a solar cell shows this phenomenon clearly from the IPCE (incident photon to current conversion efficiency) measurement data. Organic chromophore dyes of Co(PAR)(SiPA) exhibited the long wavelength absorption character compared to the N719 dye (from Dyesol).

1. Introduction

Bond-engineering as applied to semiconductor materials refers to the manipulation of the energy bands in order to control charge transfer processes in a device. When the device in question is a photocatalyst, the charges affected by drift become the focus of the study. Band-engineering is conceptually tied to defect engineering, where defect engineering aims to address issues affecting properties in the photocatalyst such as point defects, charge transfer and doping [1,2,3] General strategies are passiving co-doping to improve the photocatalys activity of semiconductors [2,4,5].

TiO_2 has several good properties, such as nontoxic, relatively inexpensive, physical and chemical stability, and high photocatalytic activity [6]. Because of these excellent properties, TiO_2 has attracted extensive attentions during decades as an extremely promising photocatalyst for environmental applications. However, TiO_2 has a wide band gap, which limits the application of visible light [6]. Therefore, the effective utilization of visible light is one of the important subjects for TiO_2 as a photocatalyst as well as a photoelectrochemical solar cell. Controled band gap engenering is immensely important to enhanced visible light absorption.

The ideal band gap of semiconductors for enhancement of photocatalyst activity can be lowered to match with visible light absorption and the location of conduction Band (CB) should be raised to meet the reducing capacity. Otherwise, by the addition of the chromophore organic dyes, the wide-band gab can be influences by interacation resulting between TiO_2 surface and the dyes. The ideal band gap for any visible light driven photocatalyst is around 2.0 eV for effective utilization of solar spectrum. The investigation of suitable semiconductor as the photocatalyst for spliting water into hydrogen using solar energy is one of the noble missions on materials science study [7-8]. In water spliting photocatalyst technology, the conduction band minimum (CBM) has to be higher (more negative) than the hydrogen reduction potential, while the valence band maximum (VBM) has to be lower (more positive) than the water oxidation potential (H_2O/O_2) [8].

Over the past two decades, dye-sensitized solar cells (DSSCs) have become a viable and relatively cheap alternative to conventional crystalline silicon-based systems. At the heart of a DSSC is a wide band gap semiconductor, typically a TiO_2 nanoparticle network, sensitized with a visible light absorbing chromophore. Ru(II)-polypyridines are often utilized as chromophores due to their chemical stability, long-lived metal-to-ligand charge transfer (MLCT) excited states, tunable redox potentials, and near perfect quantum efficiency of interfacial electron transfer (IET) into TiO_2. More recently, coordination compounds based on first row transition metals, such as Fe(II)-polypyridines and Co(II)-polypiridyl, gained some attention as potential sensitizers in DSSCs due to their low cost and abundance [9,10]

Iron(II) and Co(II) was identified many years ago as a promising substitute for ruthenium. In our study, they also have good light-absorption properties, with intense metal to ligand charge transfer (MLCT) band in visible region. While such complexes can in principle sensitize TiO_2, they do so very inefficiently since their photoactive MLCT states undergo intersystem crossing (ISC) into low-lying metal-centered states on a sub picosecond time scale [11]. Enormous efforts have been devoted to extend the light absorption range of TiO_2 from UV to visible light, such as surface modifications using chromophore dye [10]. In general, a molecul chromophore is coordinated to the surface of a wide band gap semiconductor, most commonly TiO_2. A long lived charge separation is established when the semiconductor conduction band accepts an electron from the photoexcited chromophore [6]. Molecular level factors were reported affecting the efficiency of organic chromophores for p-Type Dye Sensitized Solar Cells [12].

2. Experimental

Materials used in this research are titanium tetra isopropoxide, glass conductive indium tin oxide (ITO) (resistivity: 10 □ / cm2, Asahi Glass), triton X-100, acetylacetonates (acac), CoCl2, FeCl2, methanol, 37% hydrochloric acid, and sodium hydroxide from Merck, aminopropiltrimetosisilan (APTS) and 4- (2-piridilazoresorsinol) (PAR) from Aldrich. These materials were used directly without further purification or pretreatment particular.

2.1 Measurement electronics spectrum of 4-(2-piridilazo) resorcinol (PAR) in various pH

Measurement electronics spectrum of 4-(2-piridilazo) resorcinol (PAR) (10-3 M) was conducted at various pH condition, i.e. pH of 1, 3, 5, 6, 7, 9 and 13 in methanol. The solutions of PAR 10-3 M, in various pH were prepared by additin a certain amount of HCl solution as well as NaOH solution in order to gain the certain pH of PAR solution i.e. pH 1, 3, 5, 6, 7, 9 dan 13. Each electronic spectrum of PAR solution then were measured in the range of 200 - 800 nm using Shimadzu UV 1799 Pharma-Spec.

2.2 Formation and electronic spectrum measurement of aminopropiltrimetoksisilan (APTS), CoII-SiPA, CoII-PAR, dan SiPA-CoII-PAR

The development wide-band gap of TiO2 by the addition of organic chromophore dye, and the addition of transition metal dopand. The organic chromophore dye was synthesized by complexes compound formation of Co(PAR)(SiPA) (PAR=4-(2-piridylazoresorcinol), SiPA=Silyl propil amine). Electronic spectrum measurements of CoII-SIPA, and SIPA-CoII-PAR, was done by dissolving of CoII, and CoII-PAR with methanol into a concentration of 10-4 M (APTS at a concentration of 0.08% v/v). Each solution was measured UV-Vis absorption spectrum at a wavelength of 200-800 nm using Shimadzu UV 1799 Pharma-Spec.

2.3 Syenthesis of mesoporous TiO_2

2.993 g of Pluronic P123 was dissolved in 14 gr isopropanol (iPr) (~ 15 ml) (A solution); while we also prepared solution of 3 g TTIP + 0.53 gr acetylacetone (acac) + 4 iPr + 2.5 ml of HCl 5M (B solution) (molar ratio: TTIP / P123 / acac / HCl / H_2O / iPr = 1.00: 0, 05: 0.53: 1.25: 12.00: 30.00). The A solution was added to B solution drop wise then stirred until homogeneous for 3 hours to obtained yellow-orange sol solutions. The solution was then allowed to stand at room temperature for 7 days. The gel obtained is heated up to temperatures of 120°C for 24 hours to obtain a xerogel. After the sample allowed to stand at a temperature of ± 30°C, TiO2 xerogel was annealed at a temperature of 400°C for 2 hours with a speed temperature arise was 5°C/ min. Annnealing processing of TiO_2 mesoporous preparation was conducting using Furnace Carbolite CWF 1300.

2.4 Preparation of thin layer of TiO_2

Preparation of thin layer of TiO_2 was done by 1 mg TiO_2 was added to 1 drop of acac and 6 drops of triton X-100 and pounded until a paste TiO_2 forms. Through the Doctor Blading technique, TiO_2 paste then was coated on an ITO glass plate (1 cm x 1 cm in size). The thickness of the layer was made equal to the thickness of adhesive tape (~ 20 □m) that was used as a barrier edge. The thin layer TiO_2 then was dried in air at room temperature, and at a temperature of 400°C for 90 minutes.

2.5 Modification of wide band gap TiO2 surface structure via chromophore additon

Sensitization of TiO_2 structure was done in order to arrange sensitizer complexes in situ. TiO_2 thin layer was prepared on a conductive glass of indium tin oxide (ITO) througt slit casting. TiO_2 thin layer was dipped to a solution of 5% v/v APTS in methanol for 120 minutes for binding APTS (as an anchoring ligand) to the TiO_2 surface. After the adsorption process done completelly, the thin layer was washed with methanol and acetone, and dried at 120°C for 1 hour. CoII ions (or FeII ions) is then fastened to the anchoring ligand by means immersing thin film of modified-TiO_2 (TiO_2-SiPA) into the solution of 10^{-3} M metal ions for 120 minutes. After the immersion, the thin-layer samples were washed again with methanol, and dried at room temperature for ± 30 minutes. The last stage, the thin layer samples was dipped into secondary chromosphere of PAR (10^{-3} M) for 24 hours.

Electronic properties of TiO_2 and modified TiO_2 were done by UV Vis specular-reflectant spectroscopy (Shimadzu UV 1799 Pharma-Spec). TiO_2 thin layer and modified TiO2 (TiO2-SIPA) was characterized by X-Ray Diffraction (Shimadzu XRD-6000) on 2θ = 20-60° using CuKα radiation (λ= 1.5406 Å). X-ray diffractogram was used for determination the type of TiO_2 crystal and compared the crystal deviations before and after surface modification treatment. Silylpropilamine group attacted to TiO_2 was determined by FTIR spectrophotocopy (Shimadzu 8201 PC). Thin-layer micro-scale morphology was observed with Scanning Electron Microscope (SEM) and Energy Dispersive X-ray Analysis (EDX) equipped (Joel JSM 6360LA). By means of SEM-EDX obtained thin-layer thickness, the metals of sensitizer content have beed determined. Induced Photon to current efficiency (IPCE) of those modified TiO_2 thin layer were also calculated and compared with N719 dye modified TiO_2. IPCE spectras were measured by build DSSCs using Oriel QE/IPCE measurenment, Newport Corporation.

3. Result and Discussion

3.1. Electronic properties of 4- (2-piridilazo resorcinol) (PAR) as chromophore dye

UV-Vis absorption spectra of 4- (2-piridilazo resorcinol) (PAR) in methanol at various pH were determined using UV-Vis spectrophotometer in the wavelength range 300-600 nm (Figure 1). PAR as a ligand chromophore is expected to have four main species on the state of protonation and deprotonation is H_3PAR^+, H_2PAR, $HPAR^-$, and PAR^{2-}, as studied by Ghasemi et al. [13]. In circumstances very acidic (pH= 1), the state of the protonated H_3PAR^+ dominate the species present. The spectra are very wide of the species H_3PAR^+ likely caused by agglomeration of species H_3PAR^+ through hydrogen bridges. With the increasing of the pH system, H_2PAR species was increasing $HPAR^-$ replace apart of H_3PAR^+ species. Electronic transition of protonated species showed no separation of the electronic transition from $\pi \rightarrow \pi^*$ and $\sigma \rightarrow \pi^*$, but the electronic transition of deprotonated species, $HPAR^-$ and PAR^{2-} occured at relatively high pH showed a separations of $\sigma \rightarrow \pi^*$ and $\pi \rightarrow \pi^*$ electronic transitions (Figure 1). The $\sigma \rightarrow \pi^*$ electronic transitions have a higher energy than $\pi \rightarrow \pi^*$ electronic transition (Figure 1).

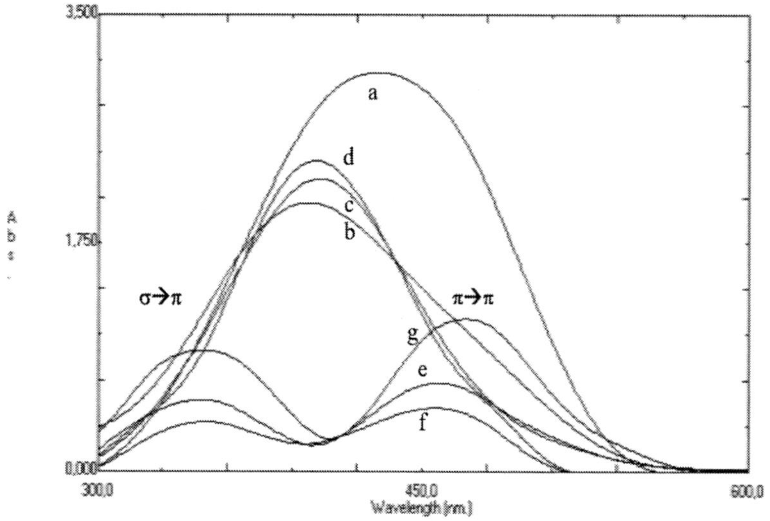

Figure 1. Electronic spectrum of 4-(2-piridilazoresorsinol) (PAR) at various pH conditions (a) pH = 1 (2) pH = 3 (c) pH = 5 (d) pH = 6 (e) pH = 7 (f) pH = 9 dan (g) pH = 13

PAR ligands have acidity constant, pKa1 = 12.08 ± 0.05, pKa2 = 5.48 ± 0.09 and 0.19 ± pKa3 = 2.98. The Ka1, Ka2 Ka3 value were governed by protonation and deprotonation equilibrium reaction of PAR. In a highly acidic condition, the dominant PAR species are H_3PAR^+, under acidic conditions (pH = 3-5) as H2PAR and at neutral and weakly alkaline (pH = 6-9) as HPAR- and PAR2- (Figure 2).

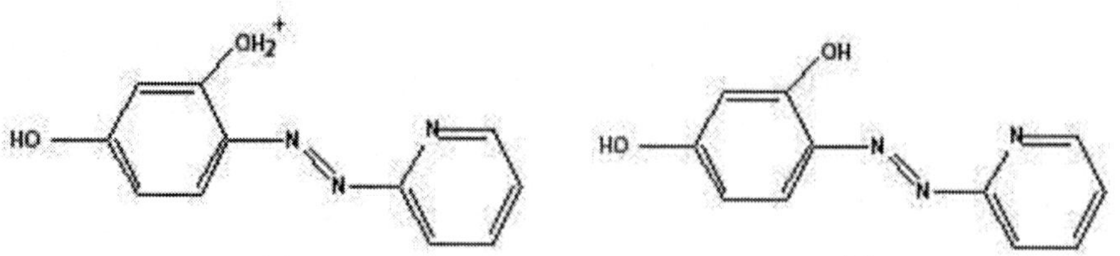

(a) (b)

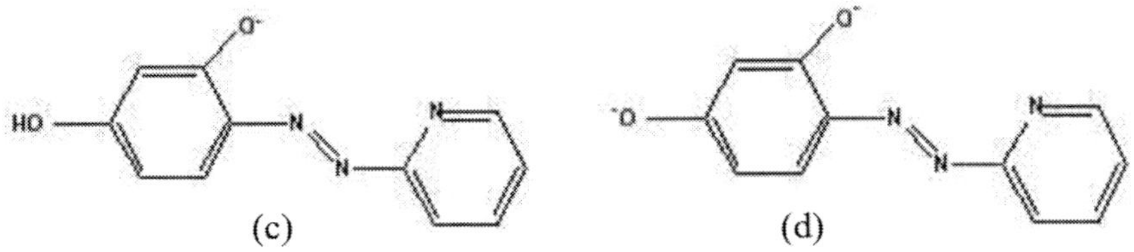

Figure 2. Mayor spesies of 4-(2-piridilazo)resorsinol (PAR) (a) H_3PAR as bidentat chromophore (b) H_2PAR as bi- and tri-dentat chromophore (c) $HPAR^-$ a a tridentat chromophore (d) PAR^{2-} as a tridentat chromophore.

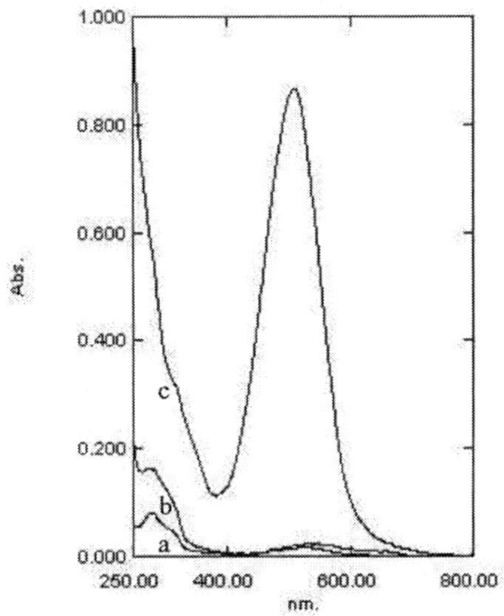

Figure 3. Electronic spectra of (a) CoII ions (b) SiPA-CoII complexes (c) SiPA-CoII-PAR complexes. The solutions were dissolved in methanol

Study of complex formation of Co^{II}-PAR describe with the electronic spectrum of Co^{II} ions, SiPA-Co^{II} complexes and SiPA-Co^{II}-PAR complexes (Figure 3). Analog with relevant pyridyl complexes compounds Rul '$(NCS)_3$ (L' = 4,4'-COOH-2,2'-bipyridine), RuL_2 $(NCS)_2$ (L = 4,4 ', 4' '- COOH -2,2 ': 6', 2 '' - terpyridine), $[Ru (bpy)_3]^{2+}$ (bpy = bipyridine)), electronic spectrum of SIPA-Co^{II}-PAR show transition electronics of intra ligand $\pi \rightarrow \pi^*$ in the UV region, the electronic transition metal to ligand charge transfer (MLCT), and d→d electronic transitions in the visible region [14-16]. In this circumstance, the electronic spectrum of SIPA-Co^{II}-PAR (Fig. 3) show electronic transitions in the visible region (λ_{maks} = 505 nm) which are transitions that generated by electronic transitions of both intra ligand $\pi \rightarrow \pi^*$PAR and ^{1}MLCT (eg → ð*) overlapping (superimposed) with d→d transitions. The electronic spectrum of SiPA-Co^{II}-PAR shows a very strong MLCT transitions in the visible region. Based on these results, the SiPA-Co^{II}-PAR complexes can be used as a photosensitizer of wide band gap inorganic oxides such as TiO_2 and ZnO. While PAR as chromophore ligand show a very strong absorption character in visible region. PAR ligand complexes (molar absorptivity = 3.27 x 10^4 M) will be able to produce a colored complexes with highly molar absorptivity.

3.2. Photosensitization process of wide band TiO2 semiconductor

In the case of photosensitizing process, the formation of complex compounds sensitizers in situ by dissolve of APTS, metal ions (Co^{II} and Fe^{II}) and PAR as starting point of sensitization process by direct methods through the creation of SiPA-Co^{II}-PAR sensitizer complexes in situ on the TiO_2 surface. APTS first attached to the TiO_2 surface, forming a bound TiO_2-SiPA. Co^{II} ion then was bounded to the surface of TiO_2-SiPA generated TiO_2-SiPA-Co^{II}; furthermore PAR will bind with TiO_2-SIPA-Co^{II} complexes resulted the TiO_2-SIPA-Co^{II}-PAR coordination.

Whereas, in other cases, sensitization of TiO2 thin layer on indium tin oxide (ITO) as a substrate through the formation of sensitizer complexes in situ is probable to permit through the reactions as shown in equation 1-3.

ITO/$\mathbf{TiO_2}$ **(1)** + $(MeO)_3$-Si-$CH_2CH_2CH_2$-NH_2 → ITO/$\mathbf{TiO_2}$≡Si-$CH_2CH_2CH_2$-NH_2 **(1)**

ITO/$\mathbf{TiO_2}$≡Si-$CH_2CH_2CH_2$-NH_2 + M(II) → ITO/$\mathbf{TiO_2}$≡Si-$CH_2CH_2CH_2$-NH_2-M^{II} **(2)**

ITO/$\mathbf{TiO_2}$≡Si-$CH_2CH_2CH_2$-NH_2-M^{II} + PAR → ITO/$\mathbf{TiO_2}$≡Si-$CH_2CH_2CH_2$-NH_2-M^{II}-PAR **(3)**

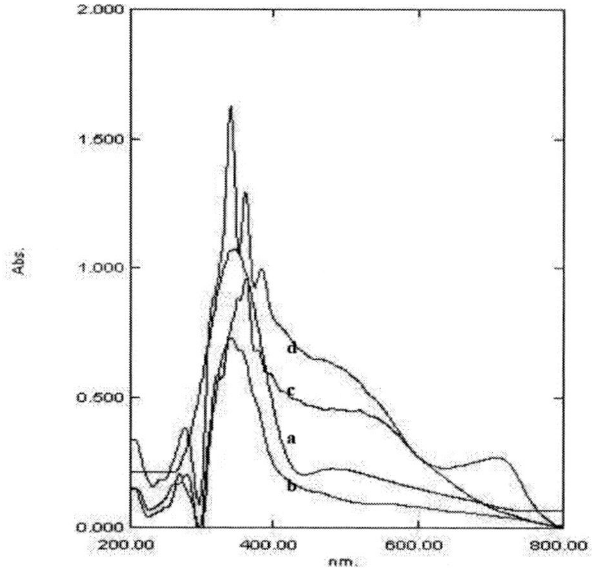

Figure 4. UV-Vis absorption spectras of (a) TiO_2 (b) TiO_2/SiPA (c) TiO_2/SiPA-Co^{II}-PAR dan (d) TiO_2/SiPA-Fe^{II}-PAR

The electronic absorption spectras in Figure 4a represents an electronic transition of TiO_2 (λ_{max}= 349 nm, λg = 405 nm). APTS has not characteristic absorption in the visible region, so that there are not absorption shift to the visible region of TiO_2 modified APTS (TiO_2-SiPA) (λ_{max} = 400 nm) (Figure 4b), while the TiO_2-SiPA-Co^{II}-PAR showed the of absorption character in the visible region (λ_{max} = 500 nm) (Figure 4c.) as $\pi \rightarrow \pi^*$ electronic transitions. This is atribute of PAR group transitions and metal-to-ligand charge transfer (1MLCT, eg →σ^*) transitions overlapping (superimposed) with d→d transitions from SiPA-Co^{II}-PAR complexes (Figure 4c). The addition of SiPA-Fe^{II}-PAR as sensitiser complexes as a comparation study in this case also show new absorption character of sensiized-TiO_2. The electronic transition at λ_{max} = 490 nm is the electronic transitions MLCT whereas electronic transition at 720 nm is a d→d electronic transition of the sensitiser complexes (Figure 4d).

The addition of APTS for surface modification of TiO_2 does not alter the crystal structure due to isomorphic substitution or transformation phase. However, analysis result of FTIR spectras show the APTS intake to the TiO_2 surfaces (Figure 5).

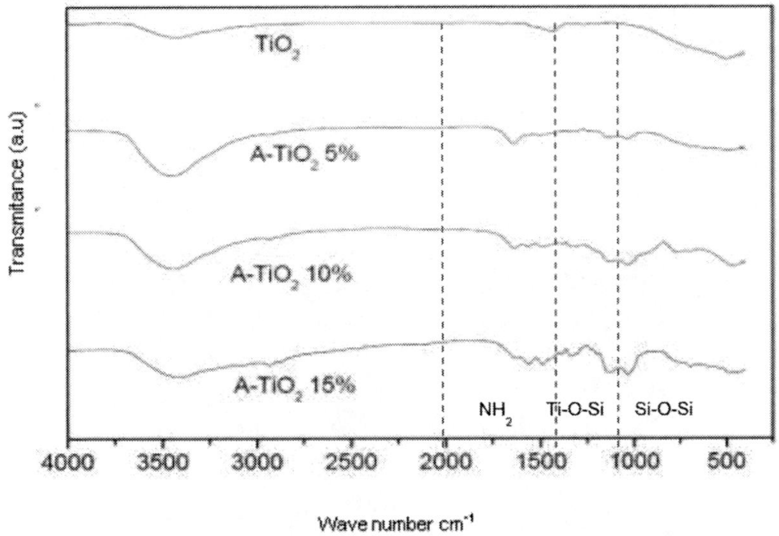

Figure 5. FTIR spectra of TiO_2 and modified TiO_2 at various APTS concentration formed TiO_2-SiPA

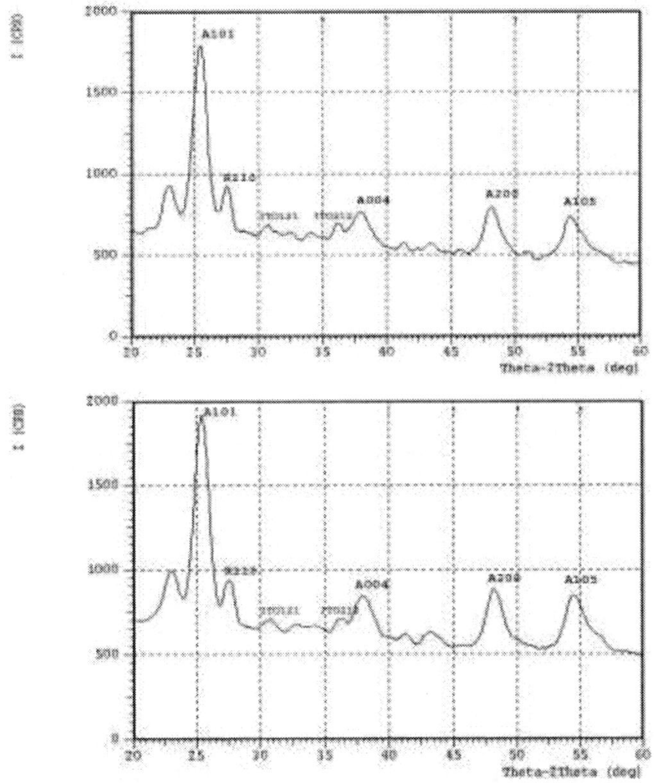

Figure 6. X-Ray difractogram of (a) ITO/TiO_2 thin film (b) ITO/TiO_2-SiPA. TiO_2 was synthesized with molar ratio of TTIP/P123/acac/HCl/H_2O/iPr = 1,00:0,051:0,53:1,25: 12,00:30,00, annealed at 400°C at a speed of 5°C/min for 2 hour. A = anatase, R = rutile, (B) = TiO2 (B)

Wide-band gap semiconductor of TiO_2 show clearly the influence of chromophore dye incoorperated in the TiO_2 surface. The results of calculations using EDX data show the presence of Co in the range of 0.08% (w/w) in ITO/TiO_2/SIPA-Co^{II}-PAR have been capable to generate the spectrum shift in the visible region. The related ITO/TiO_2/SIPA-Co^{II}-PAR thin film have morphology and thickness as shown in Figure 7. The measurement of induced photon to current efficiency of ITO/TiO_2/SiPA-Co^{II}-PAR in DSSC's shows good performance of photon to current conversion, which were depends on the wavelength absorbed by those cell (Figure 7). Chromophore group on PAR capable to absorb highly visible light. If we compare those complexes with N719 (di-tetrabutylammoniumcis-bis(isothiocyanato)bis(2,2'-bipyridyl-4,4'-dicarboxylato)ruthenium(II), the intensively light absorption is fall on the different areas. The optimum conditions of SiPA-Co^{II}-PAR light absorption are in the 600-750 nm regions.

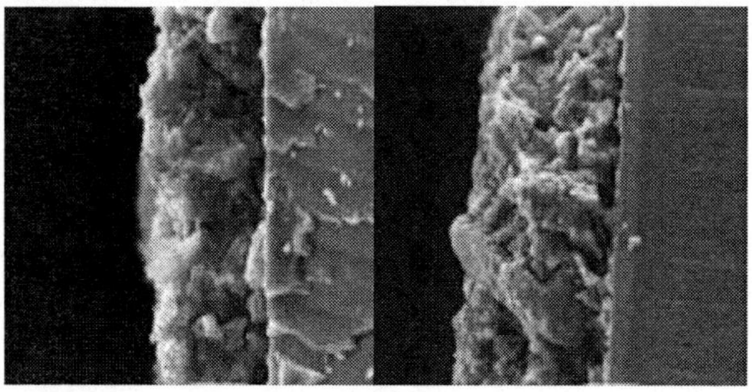

Figure 7. Micrograph SEM cross section of (a) ITO/TiO_2, (b) ITO/TiO_2-SiPA-Co^{II}-PAR. ITO/TiO_2 was annealed at 400°C for 2 hours.

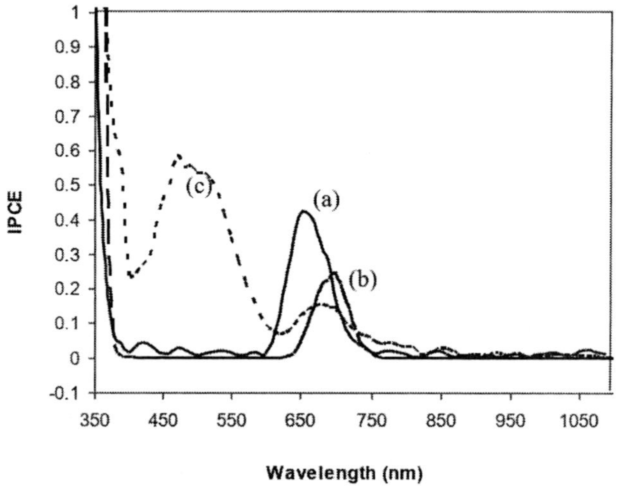

Figure 8. Normalized IPCE spectra of (a) m-TiO_2/SiPA-Co-PAR (b) P25-TiO_2/SiPA-Co-PAR. (c) m-TiO_2/N719

4. Conclusion

Photosensitizing TiO_2 with the piridilazoresorcinolcobalt(II) and piridilazoresorsinoliron(II) complexes formation in situ can improve the visible light response significantly. Improved responsiveness TiO_2 to visible light was caused by the presence of photosensitizer indicated by the appearance of the absorption spectra of sensitiser complexes of SIPA-Co^{II}-PAR and SIPA-Fe^{II}-PAR in the visible region. TiO_2 without sensitization only have an electronic transition energy of 3.55 eV

(λ_{max}= 349 nm) related to the Eg value of 3.25 eV, while TiO_2 sensitized by the SIPA-TiO_2-Co^{II}-PAR coresponded to an electronic transitions energy in visible region of 2.75 eV (λ_{max} = 505 nm) and by SIPA-Fe^{II}-PAR have electronic transitions energy in the visible region of 2.58 eV (λ_{max} = 505 nm) and 1.72 eV (λ_{max} = 720 nm), respectively. Generally, chromophore type attached to the TiO_2 surface is spesifically influence the transition energy character of sensitized-TiO_2. The IPCE (*incident photon to current conversion efficiency*) properties of the sensitized TiO_2 exhibited the long wavelength absorption character compared to the N719 dye.

References

[1] Asahi R, Morikawa T, Ohwaki T, Aoki K and Taga Y *Science* **293** 269
[2] Tian B Z, Li C Z, Gu F and Jiang H B *Catal. Commu* **10** 925–929
[3] Alivov Y, Singh V, Ding Y, Cerkovnik L J, Nagpal P *Nanoscale* **6** 10839-10849
[4] Tachikawa T and Majima T *Chem. Soc. Rev* 39 4802-4819
[5] Sun W T, Yun Y, Pan H Y, Gao X F, Chen Q and Peng L M *J. Am. Chem. Soc.* **130** 1124-1125
[6] Shane A and Gerald *J. Meyer Chem. Soc. Rev.* **38** 115-16
[7] Shahed U M K, Mofareh Al-Shahry, and William B I Jr *Science* 297 2243-2245
[8] Youngblood W J, Lee S A, Maeda K, and Mallouk T E **42** 1966-1973
[9] Yum J-H, Holcombe T W, Kim Y, Rakstys K, Moehl K, Teuscher J, J H Delcamp, Nazeeruddin M K and Grätzel M 2013 Scientific Reports **3** 2446
[10] Yella A, Lee H-W, Tsao H N, Yi C, Chandiran A K, Nazeeruddin M K, Diau E W-g, Yeh C-y, Zakeeruddin S M, Grätzel M 2011 *Science* **334** 629-634
[11] Jakubikova E and Bowman D N 2015 *Acc Chem Res.* **48** 1441-1449
[12] Karamshuk S, Caramori S, Manfredi N, Salamone M, Ruffo R, Carli S, Bignozzi C A and Abbotto A 2016 *Energies* 9 33
[13] Ghasemi J, Niazi A, Maeder M 2007 *J. Braz., Chem. Soc.* **18** 267-272
[14] Meyer T J Pure **1986** Appl Chem. **58** 1193-1206
[15] Vogler A and Nikol H 1992 *Pure & Appl.Chem* 64 1317-1317
[16] Anderson, P.A., Kenee, F.R., Meyer, T.J., Moss, J.A., Strouse, G.F., Treadway, J.A. 2002 *J.Chem. Soc. Dalton Trans* 3820-3831

Calculation of DSSC parameters based on ZnO nanorod/TiO_2 mesoporous photoanode

L Safriani[1], A Nurrida, C Mulyana, T Susilawati, A Bahtiar and A Aprilia
[1]Physics Department, Faculty of Mathematic and Natural Sciences, Universitas Padjadjaran, Jl. Raya Bandung Sumedang Km. 21, Jatinangor, Sumedang 45363, INDONESIA

E-mail: lusi.safriani@phys.unpad.ac.id

Abstract. Photoanode of dye sensitized solar cell (DSSC) plays an important role as electron transport media to accept photogenerated electron from excited state of dye. There are several physical properties that are required from photoanode of DSSC. It should be highly transparent, have large surface area, has a conduction band lower than LUMO of dye molecule, has high charge carrier mobility and finally has a good stability in redox electrolyte process. In this work, DSSC with structure FTO/ZnO nanorod/TiO_2 mesoporous/Ru-dye/gel electrolyte/ Pt/FTO has been fabricated. In order to modified the structures of photoanode, ZnO nanorod was grown on aluminium doped ZnO seed layer by variation concentration of Al (0 wt%, 0.5 wt% and 1.0 wt%). Zinc nitrate hexahydrate and hexamethylenetetramine used as raw materials for ZnO nanorod growth solution and deposited by self-assembly methods on FTO/Al doped ZnO seed layer. It is then followed by deposition of titania (TiO_2) paste by screen printing methods. DSSC parameters i.e. ideally factor (n), series resistance (R_S), and shunt resistance (R_{SH}) was derived from current density-voltage (I-V) curve using the simplify equation of ideal diode model. The influences of ZnO photoanode structures to the solar cell performance will be completely discussed.

1. Introduction

The increasing demand of energy supply has triggered many researchers to find an alternate of renewable energy resources. An interesting technology to develop is photovoltaic technology that could directly convert sunlight into electricity energy. Silicon-based semi-conductor dominated in photovoltaic applications for decades, however, the market is restricted due to their high production and environmental costs [1]. Recently, Dye-sensitized solar cells (DSSC), a low-cost photovoltaic device, have received attention to be investigated although it is required further research to enhance its efficiencies. DSSC was initiated in 1991 by O'Regan and Grätzel [2] and their devices showed efficiency as high as 12% [3]. Some advantages of DSSC are their low-cost and simple manufacturing processes, lightweight, flexible and low toxic [4].

DSSC consists of five components, a conductive substrate, a semiconductor layer, a sensitizer, a redox couple (electrolyte), and a counter electrode [5]. A thick semiconductor layer (e.g. TiO_2, ZnO and SnO_2) acted as a photoanode is coated or grown on a conductive substrate, a sensitizer (i.e. dye; e.g. N719, N3 or organic dyes or quantum dots (e.g. CdS, CdSe and PbS), an electrolyte (e.g. I_3^-/I^- and

[1] To whom any correspondence should be addressed.

Co^{2+}/Co^{3+} redox couples) injected between the sensitizer and counter electrode, and a counter electrode (e.g. Pt and carbon materials) deposited on another conductive substrate [6, 7].

One of factors that play an important role in DSSC is photoanode. Photoanode is used to collect regenerated electrons from photoexcitation of dye molecules and transfer it to external circuit through transparent electrode. On the other hand, the positive charges would be collected as cations in electrolyte side, contributed in redox cycle with incoming electrons from the external circuit to produce electrical energy [8]. Differences between TiO_2 and ZnO coated on conductive polymer as photoanode has been investigated [9]. Anatase TiO_2 nanoparticles can be produced with low cost synthetic preparation method in both academic laboratory and industry and the crystalline structure and size of TiO_2 crystals give a contribution to the efficiency of DSSC based on TiO_2 photoanodes [9,10]. Moreover, TiO_2 mesoporous as photoanode in DSSC has been reach 12% efficiency as reported by Yella et.al. [3].

Combination between ZnO and TiO_2 as photoanode is one of alternative to develop photoanode system based on metal oxide material. Mozaffari et.al reported that incorporated of ZnO nanoparticles could enhance Jsc value due to supressing of recombination reaction, which occur in TiO_2 region [11]. In this case, ZnO nanoparticles act as carrier trapping to capture charge photogenerated from TiO2/dye to reduce recombination process inside DSSC cell. In order to enhance performance of photoanode, combination between ZnO and TiO_2 based on advantages of each metal oxide. ZnO has high electron mobility (200 – 300 cm^2V/s), on the other hand TiO_2 has high energy internal surface area which could increase dye loading on photoanode system.

ZnO properties can be easily changed by engineering modification of its structure, particularly for nanorod structure. One of the ways modifications is adjusting ZnO seed layer to effect nanorod growth. The effect of seed layer condition to ZnO nanorod growth using chemical solution preparation has been investigated [12-13]. From the previous result [14], ZnO doped with a small amount of Al could enhance electrical and optical properties. Differences of Al doping concentration provide an influence on the grain growth leading to produce different surface morphology of ZnO layer. In this paper, we study the effect of Al doping concentration of ZnO seed layer (Al:ZnO) to ZnO nanorod (ZnO-Nr) growth. We also combine Al:ZnO/ZnO-Nr layer with TiO_2 mesoporous as photoanode system in quasi-solid DSSC.

The performance of DSSC can be evaluated based on sunlight-electric power conversion energy (η):

$$\eta = \frac{V_{OC} J_{SC} FF}{P_{in}} \times 100\% \qquad (1)$$

where V_{oc} is the open-circuit voltage (V), J_{sc} the short-circuit current density (mA cm^{-2}), FF the fill factor, and P_{in} the power of the incident light.

The mechanism of DSSC can be analysed through equivalent circuits, which can be useful to improve cell performance. Figure 1 shows the equivalent circuit of the single diode model which is widely used for the analysis of silicon solar cells and also DSSC [1]. The relation between external current and voltage is expressed in Equation 2.

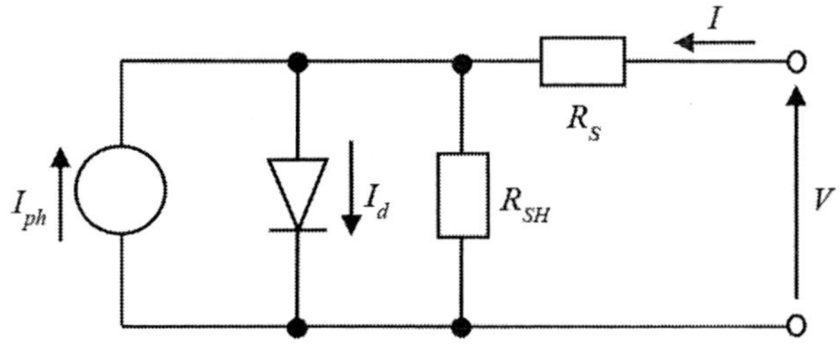

Figure 1. An equivalent circuit of single diode model

$$I = I_{ph} - I_0\left(e^{\frac{-q(V-IR_S)}{nk_BT}} - 1\right) + \frac{V-IR_S}{R_{sh}} \qquad (2)$$

where I_{ph} is photogenerated current, I_0 is dark saturation current, n is ideality factor, R_S is series resistance, R_{SH} is shunt resistance, k_B is Boltzmann's constant, q is elementary charge, and T is absolute temperature. The ideality factor n was quantitatively recognized as the inverse of the transfer factor β, and approaches to unity at high Fermi level [15].

2. Experimental

2.1. Photoanode preparation

Zinc acetate dehydrate ($Zn(CH_3COO)_2.2H_2O$) and zinc nitrate hexahydrate($Zn(NO_3)_2.6H_2O$) purchased from Merck used as raw materials for ZnO seed layer preparation and ZnO nanorods growth solution respectively. 2 methoxyethanol and diethanolamine (Sigma Aldrich) use as solvent and stabilizer on ZnO seed layer preparation. Meanwhile, alumunium chloride (Sigma Aldrich) was used as Al doping source. Deionized water and hexamethylenetetramine (HMTA) use as solvent and complexing agents on ZnO-Nr preparation.

Al:ZnO (undoped, 0.5 wt% and 1.0 wt% of Al) seed layer was deposited twice on FTO substrate using spin coater at 1500 rpm. Two step annealing treatment at 250°C and 500°C was followed to form ZnO seed layer. Self-assembly method was used to grow ZnO-Nr in isolated weighting bottle, with FTO/Al:ZnO seed layer side was facing down at 45°. Growth temperature was fixed at 100°C for 150 minutes. After that, ZnO nanorods was washed in ethanol, deionized water several times, and subsequently annealed at 500 ° C for 30 min on furnished. For photoanode system, one layer of Ti-nanooxide T/SP was deposited on FTO/Al:ZnO/ZnO-Nr by screen printing method and gradually annealed until 500°C for 30 minutes on hotplate.

2.2. DSSC assembly

FTO/Al:ZnO/ZnO-Nr/TiO_2 photoanode is ready to use as working electrode of quasi-solid DSSC. Photanode was then immersed on Ruthenium dye (535-bisTBA; purchased from Solaronix) and chenodeoxolydacid (1:10). After overnight, FTO/Al:ZnO/ZnO-Nr/TiO_2/Ru was rinsed by acetonitrile several times to remove residue of dye molecules. Drilled platinum coated FTO substrate was used as counter electrode and stack together with working electrode FTO/Al:ZnO/ZnO-Nr/TiO_2/Ru separated by hot melt surylin film (thickness 25 μm). Polymer gel electrolyte (PGE) was prepared using gel hybrid copolymer (TMSPMA:TEMS=3:1) dissolved in 0.2 ml propylene carbonate (PC) and mixed with 0.2 ml mosalyte as ionic liquid. Finally, PGE was injected to the hole of Pt/FTO side, and sealed with transparent tape to avoid leaking.

2.3. Characterizations

The scanning electron microscopy (JEOL JSM-6510A) was used to identify the surface morphology of ZnO-Nanorods. X-ray diffraction (X-pert Pro, PW3050/60 x-ray diffractometer) measurement was used to investigated ZnO-Nr structure and continued with data analysis to determine crystals lattice by Debye-Scherrer method. Current voltage (J-V) characteristics was measured using source lamp with power input 36.5 mW/cm^2 integrated with Yokogawa GS 200 DC voltage-current source and Yokogawa digital multimeter 7555.

3. Results and Discussions

The cross-sectional image of ZnO-Nr that was grown on ZnO seed layer array deposited on FTO substrate is presented in Figure 2. It is clearly shown that ZnO-Nr has already been formed and vertically aligned. SEM images of Al:ZnO/ZnO-Nr surface layer at different Al doping concentration is shown Figure 3. Diameter of ZnO-Nr is remarkably changed with the existence of Al doping in ZnO seed layer. For undoped seed layer, diameter average is around 350 nm, while for seed layer with Al doping concentration 0.5 wt% and 1.0 wt% is 120 nm and 190 nm, respectively. A small amount addition of Al doping could reduce ZnO seed layer grain size and produce more flat surface with less bulge structure due to Al^{3+} ion fill the vacant Zn sites resulted smaller crystallite size [16, 17]. ZnO-Nr that was grown on Al:ZnO seed layer showed small diameter compared to undoped seed layer. It could be explained that a nucleation sites process for nanorod growth depend on grain size of seed layer, therefore the small grain size of ZnO seed layer resulted a small diameter of nanorod [13].

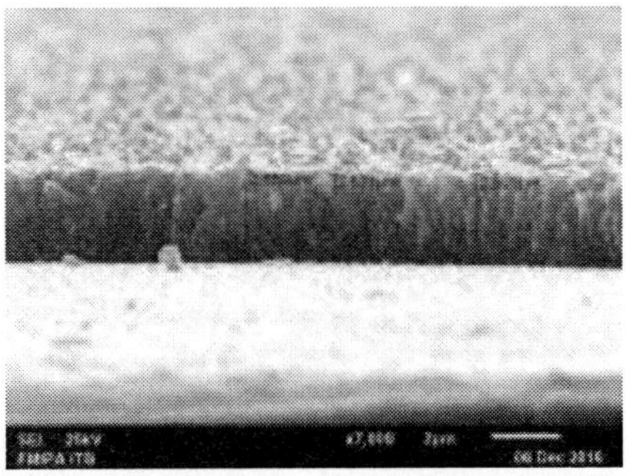

Figure 2. SEM image of cross sectional ZnO-Nrs/ZnO seed layer

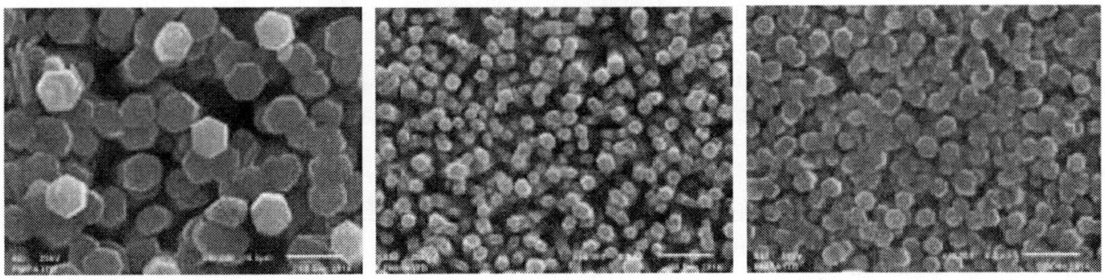

(a) (b) (c)

Figure 3. SEM images of Al:ZnO/ZnO-Nr with different Al doping concentration (a) 0wt% Al (b) 0.5wt% Al (c) 1.0wt% Al.

Figure 4 shows x-ray diffraction pattern of Al:ZnO/ZnO-Nr deposited on FTO substrate. It is clear apparently shown that peak (2θ) located between 34.3° and 34.7° corresponds with (002) plane and all samples possess pure hexagonal wurtzite structure with c-axis orientation. This result also consistent with the standard card Joint Committee on Powder Diffraction Standard (JCPDS) 36-1451. The average grain size of the ZnO-Nr was calculated using Scherrer's formula as presented in Table 1. Sample with Al concentration 1wt% has a smallest average grain size than others. The smallest peak intensity is resulted from ZnO-Nr grown on Al:ZnO seed layer with 0.5 wt% Al doping related to self-attraction between nanorod as depicted in Figure 3(b).

Table 1. X-ray Diffration parameter on Al:ZnO/ZnO-Nr with different Al doping concentration.

Crystal orientation	(002)		
Al doping concentration (wt%)	(2θ)	Crystallite grain size of ZnO-Nr (nm)	Crystal lattice c (Å)
0	34.49	25.58	5.20
0.5	34.65	21.14	5.17
1.0	34.39	15.35	5.21

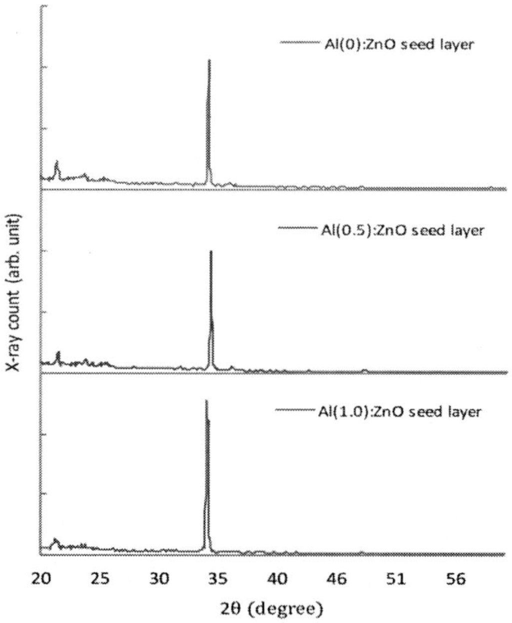

Figure 4. XRD patterns of Al:ZnO/ZnO seed layer with variation of Al concentration.

Fig. 5 shows the characteristic J-V curves for the DSSC with Al:ZnO/ZnO-Nr layer and TiO_2 mesoporous layer as photoanode. The measurement was done under illumination of 36.5 mW/cm^2 intense light. The photovoltaic parameters, which were measured, are presented in Table 2. The highest efficiency η=2.46% was obtained from device with photoanode made of Al:ZnO/ZnO-Nr with Al doping concentration 1.0 wt%, which is due to highest short current density at 3.4 mA/cm^2. A high current density indicated a higher photogenerated electron transport rate and a better interfacing between ZnO-Nr and TiO_2 mesoporous compared to other samples. As shown in SEM images (Figure 3), ZnO-Nr with Al doping concentration 1.0 wt% has more dense morphology and homogeneous therefore it gives large surface contact area with TiO_2 mesoporous and also dye particle.

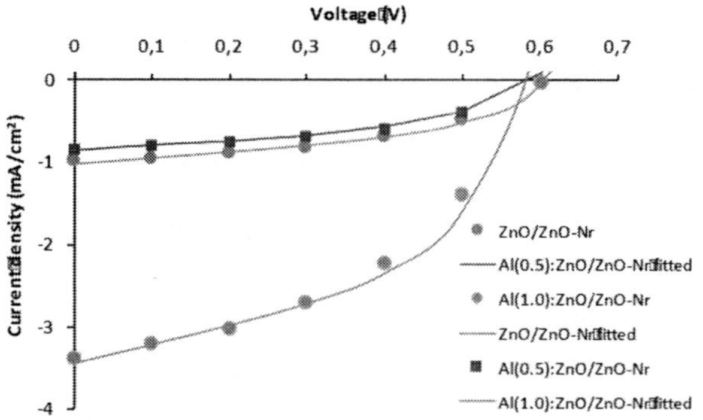

Figure 5. J-V curve of Al:ZnO/ZnO-Nr at different Al doping concentration.

Series resistance (R_S), shunt resistance (R_{SH}) and ideality factor (n) were determined using Equation 2. A raw (marked points) and fitted (solid line) data using all parameters, which obtained from Equation 2 is shown in Figure 5. The series resistance value correlated with bulk resistance of semiconductor material, metallic contact, TCO electrode, electrolyte and interconnections, and the contact resistance between the metallic contact and the semiconductor [18]. The smallest R_S was resulted from device with Al:ZnO seed layer (1.0 wt% Al doping) which lead to highest photocurrent density. A large photocurrent density also related to low recombination process inside the cell, which indicated with a small ideality factor value (n). The higher value of n could be contributed by potential drop at interfacial layer and the recombination current between metal/insulator layer [18]. In this case, employing 1.0 wt% Al doping of ZnO seed layer for ZnO-Nr growth shows smallest n value which related to low recombination process thus produce a high photogenerated current. R_{SH} inside the cell is mainly caused by leakage across the cell edge and the presence of crystal defect and/or impurities in the dye/photoanode interfacing region [18]. A high R_{SH} value indirectly could represent by high *FF* as seen on Table 2 and J-V curve (Figure 5). A deep investigation on charge transport and recombination mechanism is need to understand and explain device parameters inside the cell.

Table 2. Performance parameters of DSSC with Al:ZnO/ZnO-Nr at different Al doping concentration and TiO_2 mesoporous layer as photoanode.

Al doping concentration (wt%)	R_S (Ohm)	R_{SH} (Ohm)	n	J_{SC} (mA/cm^2)	V_{OC} (Volt)	FF (%)	η (%)
0	100	7000	3.2	1.00	0.60	46	0.76
0.5	150	14000	3.4	0.85	0.58	50	0.67
1.0	65	4500	2.9	3.40	0.60	44	2.46

4. Conclusions

ZnO-Nr has been successfully grown on ZnO seed layer doped with Al (0 wt%, 0.5 wt% and 1.0 wt%) deposited on FTO substrate using self-assembly method combined with TiO2 mesoporous layer as photoanode in DSSC. Based on SEM image and XRD measurement, a better structure and morphology of ZnO-Nr was obtained by Al:ZnO seed layer (1.0 wt% Al doping). From J-V curve, device with photoanode FTO/Al:ZnO (1.0 wt% of Al)/ZnO-Nr/TiO_2 mesoporous has efficiency 2.46% which is attributed by high photocurrent density (3.4 mA/cm^2). This work emphasizes that combination between a proper ZnO-Nr structure and TiO_2 mesoporous could enhance device performance.

Acknowledgement

This work was financially supported by Universitas Padjadjaran under research PUPT (Program Unggulan Perguruan Tinggi) with contract No. 431/UN6.3.1/PL/2016.

References

[1] Gonga J, K. Sumathya, Qiao Q and Zhou Z 2017 *Renewable Sustainable Energy Rev.* **68** 243
[2] O'Regan B and Gratzel M 1991 *Nature* **353** 737
[3] Yella A, Lee H W, Tsao H N, Yi C, Chandiran A K, Nazeeruddin M K, Diau E W G, Yeh C Y, Zakeeruddin S M and Gratzel M 2011 *Science* **334** 629
[4] Gratzel M 2006, *Prog. Photovolt: Res. Appl.* **14** 429
[5] Obotowo I N, Obot I B and Ekpe U J 2016 *J. Mol. Struct.* **1122** 80
[6] Ye M, Wen X, Wang M, Iocozzia J, Zhang N, Lin C and Lin Z 2015 *Mater. Today* **18** 155
[7] Hagfeldt A, Boschloo G, Sun L, Kloo L and Pettersson H 2010 *Chem. Rev.* **110**, 6595
[8] Basheer B, Mathew D, Benny K. George B K and Reghunadhan Nair C P 2014 *Sol. Energy* **108** 479
[9] Aravind Kumar C, Mojtaba A-J, Mohammad K N and Michael G 2014 ACS Nano **8** 2261-2268
[10] Hegazy A, Kinadjian N, Sadeghimakki B, Sivoththaman S, Allam N K and Prouzet E 2016 *Sol. Energy Mater. Sol. Cells* **153** 108
[11] Mozaffari S A, Ranjbar M, Kouhestanian E, Salar Amoli H and Armanmehr M H 2015 *Mater.*

Sci. in Semiconducting Processing **40** 285
[12] Kim K H, Utashiro K, Abe Y and Kawamura M 2014 *Materials* **7** 2522
[13] Lee H, Shin J H, Chae J, Kim J B, Kim T H and Park K B 2013 *Electron. Mater. Lett.* **9** 357
[14] Aprilia A, Wulandari P, Suendo V, Herman, Hidayat R, Fujii A and Ozaki M 2013 *Sol. Energy Mater. Sol. Cells* **111** 181
[15] Bisquert J and Mora-Sero I 2010 *J Phys Chem Lett* **1** 450
[16] Lee J H and Park B O 2003 *Thin Solid Films* **426** 94
[17] Imai Y and Watanabe A 2004 *J. Mater. Sci.: Materials in electronics* **15** 743
[18] Hafez H S, Yahia I S, Sakr G B, Abdel-Mottaleb M S A and Yakuphanoglu F 2012 *Adv. Mater. Corros.* **1** 8

Enhanced Photovoltaic Performance by Surface Modification of TiO_2 Nanorods with Aminopropyltrimethoxysilane (APTMS)

G Fadillah[1], S Wahyuningsih[1], A H Ramelan[2]

[1]Inorganic Material Research Group, Mathematic and Natural Sciences Faculty, Sebelas Maret University, Indonesia
[2]Electronic Materials and Energy, Mathematic and Natural Sciences Faculty, Sebelas Maret University, Indonesia

E-mail: sayekti@mipa.uns.ac.id

Abstract. Modification of TiO_2 nanorods through the addition of aminopropyltrimethoxysilane (APTMS) for enhancement of efficiency of solar cells has been studied. Synthesis of TiO_2 nanorods was conducted through two major stages of mechanochemical and hydrothermal refluxing at a temperature of 120 °C for 24 hours on various concentration of NaOH. Material characterizations were performed by X-Ray Diffraction (XRD), Fourrier Transform Infrared (FTIR) and Transmission Electron Microscope (TEM). Mechanochemical treatment by ball milling showed that the TiO_2 phase changed from anatase into brookite and it decreased of TiO_2 crystals size. Morphology transformation of TiO_2 to form TiO_2 nanorods was showed by rod-shaped from TEM micrographs which are characteristic of the nanorods. FTIR spectra confirmed that amine group of aminopropyltrimethoxysilane (APTMS) were successfully grafted onto the TiO_2 nanorods surface. Sensitization of TiO_2 used Ruthenium complexes N3 (N3=cis-bis(isothiocyanato) bis(2,2'-bipyridyl-4,4'-dicarboxylato ruthenium (II)) were able to increase the uptake of TiO_2 material to the visible region due to the absorption of visible light by N3 complex -APTMS. Sensitized TiO_2 nanorods were prepared for Dye Sensitized Solar Cell (DSSCs) photoanode.The maximum results of the DSSCs (Dye Sensitized Solar Cell) performance was showed that TiO_2 material modified by 10 % (v/v) APTMS capable increase efficiency of DSSCs.

1. Introduction

TiO_2 material has been widely used than the other inorganic oxide semiconductor because of its stability owned so it can be used up to several cycles. When TiO_2 particle size is reduced to the nanoscale, photovoltaic activity increases as a result of the expansion of light band-gap for quantum size and due to the enhancement of the effective surface area [1]. However, the consequences of the size of nanoscale electrodes is slow electrons movement because of its through a connection between nano particles randomly distributed [2].

In recent years, several studies reported that promising potensial of change TiO_2 three-dimensional (3-D) into one-dimensional (1-D) such as nanorods, nanotubes and nanowires have received considirable attention due to their unique properties and novel applications [3, 4]. One of the most important and unique properties of TiO_2 1-D is that the transfer of electron and holes in semiconducting materials is mainly governed by the quantum confinement phenomenon and transport characteristics related to photons and photons are largely affected by the size and geometrical building of the materials [2, 5]. Besides that, modification of nanostructure TiO_2 into 1-D is able to provide increase surface area of TiO_2 [1]. However, the major problem of application TiO_2 nanorods in photovoltaics is dye adsorption onto TiO_2 nanorods. Dye adsorption onto TiO_2 is highly challenging due to the lack of chemical bonding between semiconductor TiO_2 and dye [6]. On the other hand, grafting active functional groups onto particle surface can increase the possibility of chemical bonds formation between TiO_2 and dye. Chen and Yakovlev (2010) [7] investigated the adsorption and interaction of organosilanes on TiO_2 nanoparticles and confirmed the bonding of functional groups on particle surface was realized through Si-O-Ti bonds.

In this study, the types of silane coupling agent aminopropyltrimethoxysilane (APTMS) was applied as a particle surface modifier for grafting functional groups onto surface of TiO_2 nanorods. Modification of semiconductor TiO_2 is efforts to increase the effectiveness of photovoltaic semiconductor with the aim of preventing the recombination of electrons (hereinafter abbreviated as e^-) and holes (hereinafter abbreviated h^+) to increase the conversion Incident Photon to Current Efficiency (% IPCE). In order to understand the effect of surface modifications on the TiO_2 nanorods photovoltaics activity was investigated.

2. Experimental

2.1. Materials

TiO_2 powder was provided from commercial TiO_2. Aminopropyltrimethoxysilane (APTMS) was purchased from ABCR, Germany. Sodium hydroxide (NaOH) and ethanol. was provide by Merck. The material was used as received without any pretreatment.

2.2. Synthesis of TiO_2 nanorods

The ratio of ball :TiO_2 = 20:1 was inserted into a planetary ball milling at the rotation speed of 1000 rpm for 4 hours. 3 g of TiO2 from the milling process was then added to aqueous NaOH solutions at various consentration of 8, 10 and 12 M. The mixture was then refluxed for 24 hours at a temperature of 120 °C. The mixture was neutralized with 0.1 M HCl until reach pH 7. The mixture was filtered and dried at a temperature of 60 °C for 12 hours and calcined at 400 °C for 2 hours.

2.3. Surface modification of TiO_2 nanorods

0.5 g of TiO_2 nanopowder, was dispersed in 2 ml ethanol to obtain a pasta. Then, the pasta was coated on FTO (Flourine doped Tin Oxide) glass with slip casting method. Then, the silane coupling agents of APTMS with different concentrations (5, 10 and 15%) were added in the thin layer of TiO_2 and allowed for 1 minute. After that, the the thin layer of TiO2 was washed with ethanol and water for at least 2 cycles to remove excessive silanes. The modified particles as thin film FTO/TiO2-APTMS were dried in an oven at 40 oC for 30 minutes

2.4. Photovoltaics characteristic performance

Thin layer of FTO/TiO_2- APTMS was soaked in a dye solution of N3 Ruthenium complexes (3 x 10-4 M) for 24 hours. Whereas, a counter electrode was prepared by addition of carbon powder onto FTO glass. Both electrodes were sealed and given a distance to place electrolyte solution. After the electrolyte was inserted, the photoanoda cell was tested by a IV Keithley 2602A measurment by cell illumination cell by a halogen lamp (1000 W/m2).

3. Result and Discussion

3.1. Synthesis of TiO_2 nanorods

A material TiO_2 nanorod has been synthesized from TiO_2 nanoparticle that has been treated ball milling with the rotation speed of 1000 rpm for 4 hours. Figure 1 shows the results of XRD pattern of TiO_2 before and after ball milled treatment.Transformation of the X-ray diffraction pattern (Fig. 1) shows the revolution in anatase phase TiO_2 into brookite phase TiO_2. This mechanochemical technique of TiO_2 preparation has also been carried out by Dutta *et al* (2002) [8] and Rezaee *et al* (2011) [9]. Dutta *et al* (2002) have reported the phase transformation of TiO_2 crystals that can occur through mechanochemical ball milling treatment [8]. Those processes were influenced by the rotational speed and agitation time.

IOP Conf. Series: Earth and Environmental Science 75 (2017) 012005 doi:10.1088/1755-1315/75/1/012005

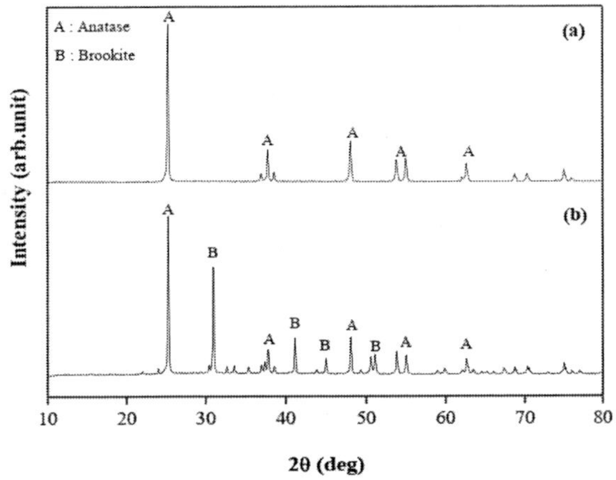

Figure 1. X-ray diffraction pattern of TiO_2 (a) before ball milled process (b) after ball milled process

We used ball milled TiO_2 to prepare TiO_2 NRs. Synthesis of TiO_2 NRs has been done by hydrothermal refluxing method at a temperature of 120°C using a strong base of NaOH on various concentration of 8, 10, and 12 M for 24 hours. The hydrothermal process can be divided into four stages, e.g 1) the synthesis of nano-TiO_2 in an alkaline solution, 2) alkaline hydrolysis through bond disconnection Ti-O-Ti, 3) re-polymerisation by condensation process 4) crystallization growth through annealing process. If TiO_2 is reacted with a strong base it will hydrolyse, this is splitting the polymer chains of TiO_2 (depolymerization) by the presence of OH^- groups. Depolymerization occurs because the strength of OH^- ions that can break the Ti-O-Ti in the polymer. A result of subsequent alkaline hydrolysis was neutralized by the addition of HCl. At pH 7, it will obtain the most optimal condensation process. Condensation processes are H_2O releasing at the two ends of the polymer chain due to polymerization process. The condensation material result was heated at a temperature of 400 °C to obtain TiO_2 nanorods. TiO_2 after calcination at 400 °C were characterized using XRD shown in Figure 2. NaOH that used to synthesize TiO_2 nanorods influence the TiO_2 structure. XRD pattern of TiO_2 nanorods (Figure 2) shows that anatase TiO_2 phase is as most abundance. Rutile phase TiO_2 was increasing when the NaOH was added about 12 M. These results are related to the synthesis results reported by Tsai and Teng (2006) that the nano TiO_2 powder can be synthesized from a mixture of rutile and anatase phase at various concentration of NaOH [10].

Physicochemical treatment of TiO_2 influence both phase and size of TiO_2 crystal. Calculation using Scherrer equation resulted that the ball milled TiO_2 composed by anatase TiO_2 54.07 nm in crystal size, and brokite TiO_2 41.19 nm in crystal size. While after hydrothermal refluxing process, the crystal size of TiO_2 decrease into 7.65 nm. Gribb and Banfield (1997) have described that tendencious anatase phase transform to more stable phase of rutile if the crystal size extent to about 30 nm [11]. But the brokite phase was rare occur because of lower stability properties. Brokite phase in most case will develop at annealed in a range temperature of 200 - 500 °C [12]. TiO_2 annealed at a higher than 500 °C shows the lower anatase as well as brokite content, finally no found both of them at annealed result at more higher than 800 °C [13].

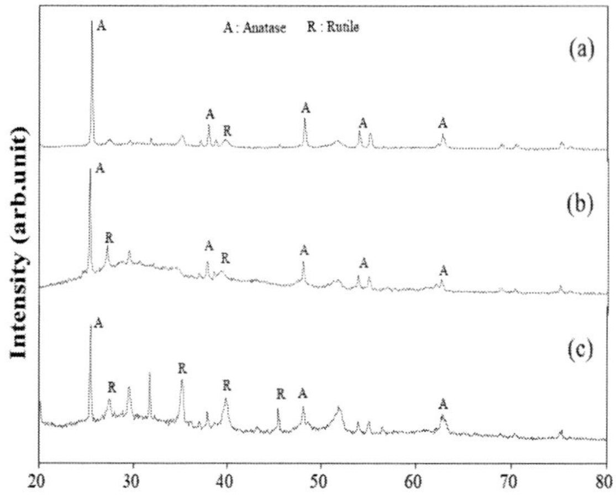

Figure 2. XRD pattern of TiO_2 nanorods treated by NaOH solution at (a) 8 M, (b) 10 M, dan (c) 12 M.

Representative transmission electron microscopy (TEM) images of treated TiO_2 under NaOH solution are provided in Figure 3. An overview of the nanorods at low magnification in Figure 3c shows that the sample almost exclusively contains nanorods with an average length of 80 nm and diameters of 50 nm. But in the lower NaOH concentrations (Figure 3a and Figure 3b) 1D nanorods not appear.

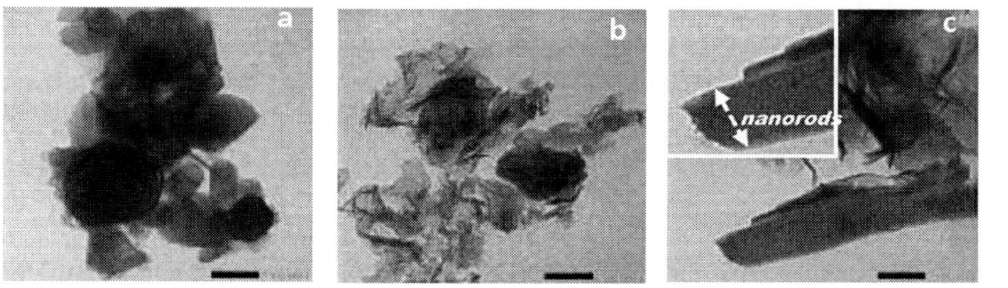

Figure 3. TEM micrographs of the TiO_2 treated using NaOH solution of (a) 8 M, (b) 10 M, dan (c) 12 M

TiO_2 preparation results using NaOH 10 M , showed flaking 3D structure, then has formed TiO_2 2D structure (Figure 3b). We further discoverred 1D structure of TiO_2 in preparation results using NaOH 12 M. Structural transformation of the ball milled TiO_2 to 1D TiO_2 nanorods was characterized from pore size distribution data. 1D TiO_2 nanorods have an average surface area 79 m^2/g, but ball milled TiO_2 only have it about 7.56 m^2/g. It was concluded that structural transformation to 1D nanorods capable increase surface area. Increasing of surface area will be increase dye adsorption onto TiO_2 surface. It will gain good performa for dye sensitized TiO_2 photoanode.

3.2. Modified of TiO_2 nanorods

The modification process of APTMS on the TiO_2 nanorods surface is illustrated in Figure 4(2). Different concentrations of silane coupling agents to TiO_2 have a significant influence on FTIR spectra. As can be seen in Figure 4, the transmission FTIR spectra of TiO_2, TiO_2-APTMS 5% , TiO_2-APTMS 10%, and TiO_2-APTMS 15% were shown. From spectra of TiO_2-APTMS the peaks below 700 cm^{-1}, which were assigned to Ti-O and Ti-O-Ti bonding of titania were ignored in this case because of their over saturated absorption [7]. The stretching vibration of absorbed water as well as surface hydroxyl groups (OH) which presented in the TiO_2 were confirmed by the broad absorption band between 3400 and 3200 cm^{-1} and the low intensity peak at 1640 cm^{-1} [7]. After surface modification by APTMS, as presented in spectra of TiO_2-APTMS 5 % , TiO_2-APTMS 10 %, and TiO_2-APTMS 15 %, the asymmetrical and symmetrical stretching vibration of the C-H bond in methylene group were observed at 2928 and 2870 cm^{-1} [14]. Furthermore, the peak corresponding to Si-O-Si bond was observed at around 1040 cm^{-1} indicating the condensation reaction between silanol groups. The N-H bending vibration of primary amines ($-NH_2$) was observed as a broad band in the region 1605–1560 cm^{-1}, and another low intensity peak on the shoulder of titania peak at 1140 cm^{-1} was assigned to the C-N bond. The appearance of these bands demonstrated that amine functional groups in organosilane were grafted onto the modified surface.

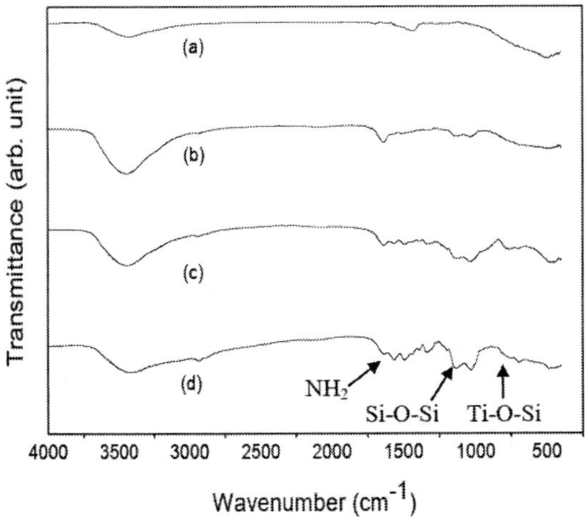

I

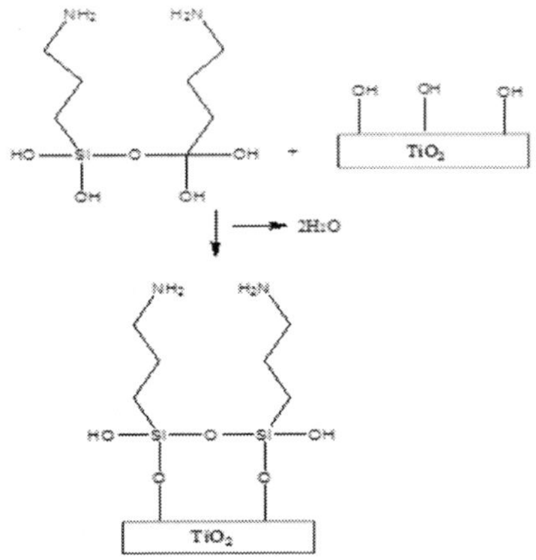

II

Figure 4. (1)FTIR spectra for (a) TiO_2 (b) TiO_2-APTMS 5 % (c) TiO_2-APTMS 10 % (d) TiO_2-APTMS 15% ; (2) Illustration of surface modification TiO_2 by APTMS

3.3. Photoelectrochemical solar cells perform of modified TiO_2 NRs

Modification of surface TiO_2 nanorods with APTMS was able to increase adsorption N3 complexes. APTMS has a amine group with a lone pair electrons which is caused the chemical bond between TiO_2 and N3 complexes more easy. Amount of the dyes was higher adsorbing onto modified TiO_2 compared with it onto unmodified TiO_2 NRs.

Figure 5 shows the current density-voltage curves of the dye sensitized solar cells based on the TiO_2 nanorods without modification and TiO_2 with modification on various concentration of APTMS (5,10 and 15 % (v/v)) composites. The values of short-circuit current density (J_{sc}), open-circuit voltage (V_{oc}) and overall power conversion efficiency (η) are summarized in Table 1.

Table 1. Performance of dye sensitized solar cell using unmodified and modified TiO_2 NRs thin film photoanodes

Sample	V_{oc} (Volt)	I_{sc} (mA)	P_{max} (mW)	Efficiency, η (%)
TiO_2	0,00009	0,00004	0,0046	0,0023
TiO_2-APTMS 5 %	0,199	0,016	0,750	0,370
TiO_2-APTMS 10 %	0,400	0,015	2,260	1,130
TiO_2-APTMS 15 %	0,555	0,005	0,960	0,480

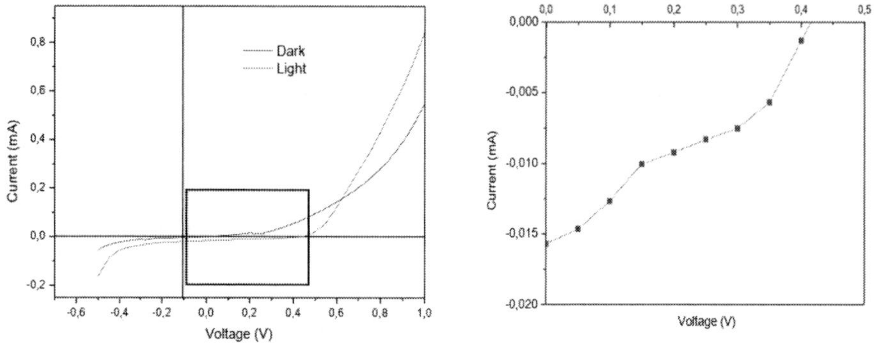

Figure 5. Current density-voltage curves of the dye sensitized solar cells based on the TiO_2 nanorods with modification of APTMS 10 %.

Overall, photovoltaic performance test was showed that modification of TiO_2 nanorods can increase efficiency of solar cell. Howover, modification of APTMS on concentration 15 % doesn't provide optimum efficiency. Modification of high concentration was likely to occur polymerization from APTMS, which the polimerization will increase the resistivity of material [15]. Polymerization was triggered by bond formation at the end of the chain, it is causing hole recombination and the efficiency will be lower [16].

4. Conclusion

In this study, the effect of surface modification TiO_2 nanorods with silane coupling agent (APTMS) was investigated. The characterization using TEM and FTIR confirmed that surface modification has been successfully grafted onto TiO_2 surface through Ti-O-Si chemical bonds. By the modification of APTMS 10 % (v/v) on the surface TiO_2 nanorods have been capable improved cell performance which was efficiency of 1.13 %.

References

[1] Karimi L, Mirjalili M, Yazdanshenas M E and Nazari A 2010 *Photochem. Photobiol* **86** 1030 1037
[2] Mohamed A L R and Rohani S 2011 *Energy and Envirotmental Science* **4** 1065
[3] Pan Z W, Dai Z R and Wang Z L 2001 *Science* **291** 1947 – 1949
[4] Patzke G R, Krumeich F and Nesper R 2002 *Oxidic Nanotubes and Nanorods-Anisotropic Modules For a Future Nanotechnology* (Switzerland : Angew Chem. Int.)
[5] Wang Y Q, Hu G Q, Duan X F, Sun H L and Xue Q K 2002 *Chemical Physics Letters*. **365** 427 – 431
[6] Sojka-Ledakowicz J, Lewartowska J, Kudzin M, Leonowicz M, Jesionowski T, Siwinska-Stefanska K and Krysztafkiewicz A 2009 *J. Mater. Sci.* **44** 3852 - 3860
[7] Chen Q and Yakovlev N L 2010 *Appl. Surf. Sci.* **257** 1395 – 1400
[8] Dutta H, Sahu P, Pardhan S K and De M 2002 *Material Chemistry and Physics* **77**153 - 164
[9] Rezaee M, Khoie S M M and Liu K H 2011 *Cryst Eng Comm* **13** 5055 – 5061
[10] Tsai C C and Teng H 2006 *Chem. Mater* **18** 367 – 373
[11] Gribb A and Banfield J F 1997 *American Mineralogist* **82** 717 – 728
[12] Bakardjieva S, Stengl V, Szatmary L, Subrt J, Lukac J, Murafa N, Niznansky D, Cizek K, Jirkovsky J and Petrova N *2006 J. Mater. Chem* **16** 1709 - 1716
[13] Wahyuningsih S, Fadilah G, Hidayat R and Ramelan AH 2016 *Procedia Chemsitry* **19** 632-637
[14] Ukaji E, Furusawa T, Sato M and Suzuki N 2007 *Appl. Surf. Sci.* **254** 563 – 569
[15] Ogasawara T, Yoshino A, Okabayashi H and O'Connor C J 2001 *Colloids and Surfaces A: Physicochem. Eng. Aspects* **180** 317 – 322
[16] Zhao J, Milanova M and Marijn M C G 2012 *Colloids and Surfaces A: Physicochem. Eng. Aspects*. **413** 273 - 279

IOP Conf. Series: Earth and Environmental Science **75** (2017) 012006 doi:10.1088/1755-1315/75/1/012006

Synthesis of TiO_2 NRs - ZnO Composite for Dye Sensitized Solar Cell Photoanodes

S Wahyuningsih[1], A H Ramelan,[2] R Hidayat [1], G Fadillah[1], H Munawaroh[1], L N M Z Saputri[1]
[1] Inorganic Material Research Group, Faculty of Mathematics and Natural Sciences, Sebelas Maret University, Ir. Sutami street 36A Kentingan Surakarta Indonesia
[2] Electronic Materials and Energy, Faculty of Mathematics and Natural Sciences, Sebelas Maret University, , Ir. Sutami street 36A Kentingan Surakarta Indonesia

E-mail: sayekti@mipa.uns.ac.id

Abstract. Composite of TiO_2 NRs - ZnO were synthesized for DSSCs photoanode materials. TiO_2 NRs was synthesized from TiO_2 anatase by mechanochemical technique using ball milling process with agitation speed of 1000 rpm. While, the further hydrothermal refluxing process was conducted at 120°C under various concentration of NaOH in aqueous solution. The starting material of ZnO was prepared from $ZnSO_4.7H_2O$ as a precursor. The hydrothermal treated TiO_2 was added to the ZnO powder in a certain composition of 1:1, 1:2 and 2:1 (w/w), and the mixtures were then annealed at 400°C. The resulting material was characterized by X-ray diffraction (XRD), Surface area analyzer (SAA), Transmission electron microscopy (TEM), and Thermogravimetry/Differential thermal analysis (TG/DTA). The TiO_2 revolution occurs from anatase phase into brookite phase. Rutile TiO_2 phase was increasing when the NaOH was added at about 12 M. Nanograf of TEM showed the optimum condition for the formation of TiO_2 NRs was obtained when 12 M NaOH was used. Structural transformation to 1D nanorods of TiO_2 capable increase surface area up to 79 m^2/g. TiO_2 NRs–ZnO composite was prepared from TiO_2 NRs and ZnO using comparation of TiO_2 NRs: ZnO = 1:1, 1:2, dan 2:1. Anatase phase TiO_2 as a single phase TiO_2 was obtained in the TiO_2–ZnO composite (1:1 w/w) upon heating the sample until 400°C. Difference TiO_2 NRs-ZnO composite materials were investigated as good photovoltaic materials. Evaluation of the performance of DSSCs was conducted by I-V Keithley 2602A measurement indicate that photoanode built of TiO_2 NRs - ZnO thin film has a higher solar cell efficiency than that of TiO_2 thin film photoanode.

1. Introduction

Nanomaterials are tremendous interest due to their noticeable application in electronics, optics, and photonics as well as photocatalytic materials. Zinc oxide as a direct wide-band gap semiconductor with a bandgap of 3.37 eV and a large exciton binding energy of 60 meV at room temperature is discovered to be importance semiconductor photocatalysis in recent years. In general, the photocatalytic activity is based on the generation of electron/hole pairs and limited by the recombination of electron/hole pairs. Many methods, such as doping, metal modification, and coupling of composite semiconductors, have investigated to suppress the recombination and increase the life time of separation electron and hole, hence to improve the photocatalytic properties. Among them the coupling of different semiconductor seems useful to achieve the recombination of electron/hole separation and prevent the recombination electron hole pairs efficiently [1]. Difference TiO_2/ZnO composite materials were investigated as a good photocatalytic as well as a good photovoltaic materials.

Different author apply various methods of modification like ball milling [2], doping and codoping [3, 4] to obtain efficient photocatalysis under visible light with ZnO, TiO_2 and TiO_2-ZnO composites. Application of supported bycomponent catalysts from two semiconductors oxide also alter the electronic properties, as compared to the initial materials, in turn affecting charge photogeneration, charge separation and transfer, as found for ZnO overlayers on TiO_2. Although researcher focused on the fabrication of TiO_2/ZnO composite materials, and their photoactivity, there is still a need to study the structure and growth properties influence on the photocatalytic as well as photovoltaic activity.

Jayasankar et al. (2012) [5] have extend the hybrid sol–gel–powder method to the processing of thick, macro-meso-porous TiO_2, ZnO and ZnO–TiO_2 composite films, and show that these films can be advantageously applied for the photocatalytic degradation of dye molecules, as one of possible applications. The results presented hereafter show a procedure for mixing TiO_2 and ZnO nanopowders with optimized TiO_2 and ZnO precursor solutions, respectively, to process thick films with homogeneously distributed nanoparticles and porosity. ZnO–TiO_2 composite thick films were also investigated. The films were processed via spin coating and annealed at a fairly low temperature of 400°C to yield approximately 4 mm thick films in one coating sequence. Over the past several years, we have developed ZnO nanorod arrays suitable for both polymer-inorganic cells and dye-sensitized cells. Several groups have built polymer-inorganic cells from their own ZnO rod arrays, but in general the devices perform [6].

Nanoparticles (NPs) decorated ZnO/TiO_2 core/shell nanorod arrays were fabricated on transparent conductive glass substrates by sequential plasma deposition and post-annealing processes for dye-sensitized solar cells (DSSCs) applications. The NPs decorated ZnO/TiO_2 nanorods were composed of single-crystalline ZnO nanorods, homogeneously coated thin TiO_2 shells and entirely covered anatase TiO_2 NPs. The photocurrent density of the composite electrode was largely enhanced due to the enlarged surface area, the dark current was suppressed and the open-circuit voltage was increased because of the energy barrier formed at the interface between the ZnO core and the TiO_2 shell [7].

DSSCs based on one-dimensional (1D) ZnO nanostructures, which exhibit significantly higher electron mobility than that of both TiO and ZnO NPs films [8], have recently been attracting increasing attention [8, 9]. However, the power conversion efficiency of these DSSCs based on ZnO nanorods was still at a relatively low level (~1.5%) [8]. Insufficient internal surface area of those 1D nanostructures is considered to be one of the key factors in limiting the energy conversion efficiency. For improving the surface area without sacrificing electron transport efficiency, nanorods/nanoparticle composite films [7], were employed to be used as the photoanode of DSSCs. The photocurrent density and efficiency of these cells were largely increased because of the enriched surface area for better dye loading and lightharvesting.

2. Experimental

2.1. Synthesis of TiO_2 NRs –ZnO

Treated TiO_2 ball mill was refluxed under NaOH solution of 8, 10, and 12M for 24 hours. Then the powder resulted was filtered and neutralized to pH 7 using 0.1 M HCl to obtain a solid after dried at 60°C for 12 hours. While, the ZnO nanoparticle was synthesized by $ZnSO_4.7H_2O$ precursor using NaOH alcoholic solution. Those precursor was also refluxed for 2 hours at 60°C until ZnO powder formed. Then the solids obtained were dried at a temperature of 60 - 70°C. Composites of TiO_2 NRs - ZnO were prepared at ratio of TiO_2 NRs : ZnO = 1: 1, 1: 2 and 2: 1 (w/w) in ethanol, respectively. Then, it was stirred for 10 minutes at room temperature. TiO_2 NR_S - ZnO composites were annealed at a temperature of 400°C. the TiO_2 NRs-ZnO was characteried by XRD, TG/DTA, SEM and TEM.

2.2. Fabrication of DSSCs

DSSCs fabrication used FTO glass as a substrate. Nanorods composite of TiO_2 –ZnO at various compositions as about 5 grams was dissolved in 2 mL of ethanol to obtain a paste. The pasta was then coated on FTO glass with slip coating method. Pt paste was also superimposed on another FTO glass to built counter electrode thin film. Both FTO based electrode was heated at a temperature of 400°C for 15 minutes. Composite materials photoanoda was soaked in a dye solution of N3 ruthenium complexes for 24 hours. Then, both electrodes were sealed and given a distance to immerse the electrolyte solution. Electrolyte was prepared from potassium iodide (KI) of 0.8 grams and iodine (I_2) of 0.127 grams dissolved in 10 mL acetonitrile. After the electrolyte was immersed, the DSSCs was characterized by an I–V Keithley 2602A measurenment.

3. Result and Discussion

3.1. Synthesis of TiO_2 NRs –ZnO

Material TiO_2 nanorods (NRS) has been synthesized from TiO_2 nanoparticle that has been treated ball milling with agitation speed of 1000 rpm. Figure 1 shows the results of X-ray diffraction characterization of TiO_2 before and after ball milled treatment. Transformation of the X-ray diffraction pattern (Figure 1) shows the revolution in anatase phase TiO_2 into brookite phase TiO_2. This mechanical-chemical technique of TiO_2 preparation has also been carried out by Dutta et al. (2002) [10] and Rezaee et al. (2011) [11]. Dutta et al (2002) [10] have reported the phase transformation of TiO_2 crystals that can occur through mechanical-chemical ball milling treatment. Those processes were influenced by the rotational speed and agitation time.

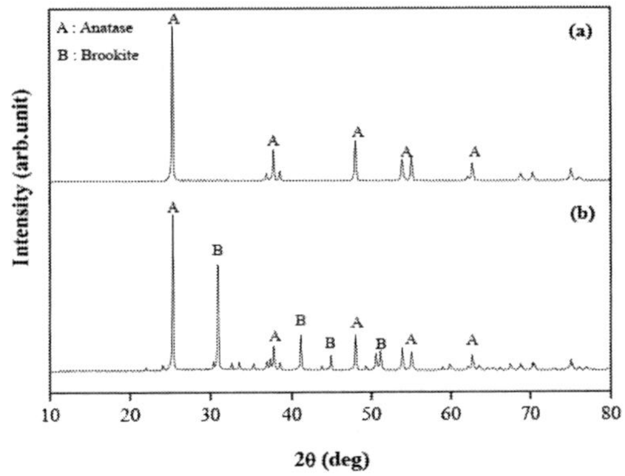

Figure 1. X-ray diffraction pattern of TiO_2 (a) before ball milled process (b) after ball milled process

We used ball milled TiO_2 for preparation of TiO_2 NRs. Synthesis of TiO_2 NRs has been done by hydrothermal refluxing method at a temperature of 120 °C using a strong base of NaOH at a concentration of 8, 10, and 12 M for 24 hours. TiO_2 after calcination at 400 °C were characterized using XRD shown in Figure. 2.

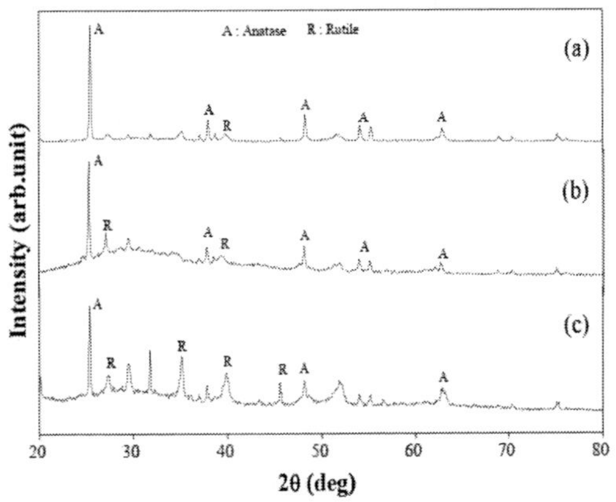

Figure 2. XRD pattern of TiO_2 NRs treated by NaOH solution at (a) 8M, (b) 10M, and (c) 12M

The NaOH solution used to synthesize TiO_2 NRs influence the TiO_2 structure. XRD pattern of TiO_2 NRs (Figure 2) shows the most abundance of anatase TiO_2 phase. Rutile TiO_2 phase was increasing when the NaOH was added at about 12 M. These results are related to the synthesis results reported by Tsai and Teng (2006) [12] that the nano TiO_2 powder can be synthesized from a mixture of rutile and anatase phase at various concentration of NaOH. Representative transmission electron microscopy (TEM) images of treated TiO_2 under NaOH solution are provided in Figure 3. An overview of the nanorods at low magnification in Figure 4c shows that the sample almost exclusively contains nanorods with an average length of 80 nm and diameters of 50 nm. But in the lower NaOH concentrations (Figure 3a and Figure 3b) 1D nanorods weren't appear.

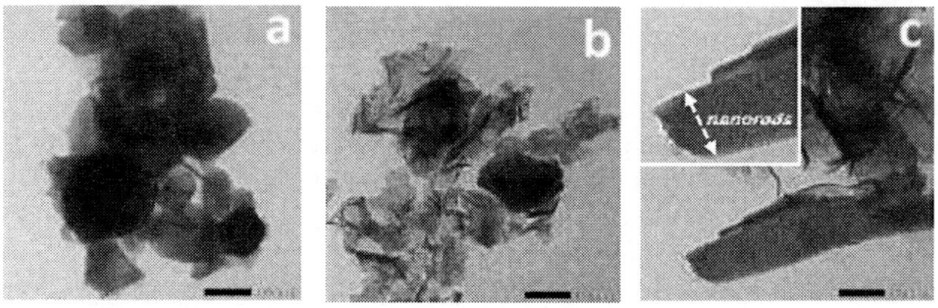

Figure 3. TEM micrographs of the TiO_2 treated using NaOH solution of (a) 8 M, (b) 10 M, dan (c) 12 M

Structural transformation of the ball milled TiO_2 to 1D TiO_2 NRs was characterized from pore size distribution data's (Figure 4). 1D TiO_2 NR*s* have an average surface area 79 m^2/g, but ball milled TiO_2 only have it about 7.56 m^2/g. It was concluded that structural transformation to 1D nanorods capable increase the surface area. Increasing of surface area characteristic will be increase dye adsorption onto TiO_2 surface. It will gain well perform for dye sensitized TiO_2 photoanode.

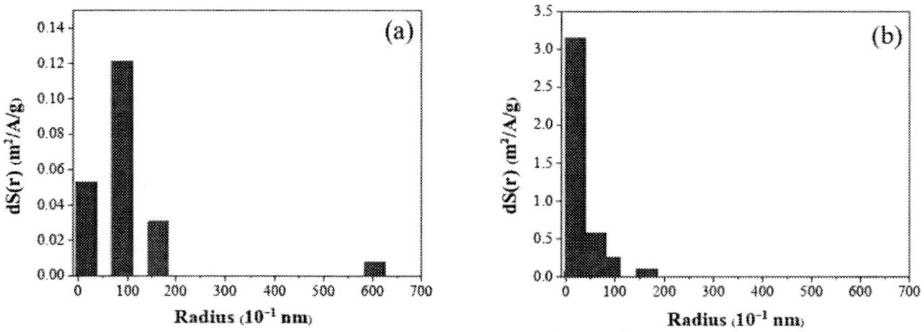

Figure 4. Pore size distribution of (a) ball milled anatase TiO_2 nanoparticle (b) TiO_2 NRs

TiO_2 NRs–ZnO composite was prepared from TiO_2 NRs and ZnO using molar ratio of TiO_2 NRs : ZnO = 1:1, 1:2, and 2:1, respectively. Thermogravimetry and Diferential Thermal Analysis (TG/DTA) (Figure. 5) show the crystallization process in the range temperature of 30 – 900°C. Mass decrease in the range of 60 – 150°C and 200 – 350°C can be ascribed to release H_2O content dan decomposition of organic volatile compound (Figure 5a). DTA/TGA curves showed significant differences of each sample (Figure 5). TiO_2 NRs (sample A) showed a shoulder exothermic peak at 250°C and a sharp and strong exothermic peak at 360°C, and at 700°C, they can be ascribed to dehydroxylation, anatase and rutile formation, respectively. It seems that those process not similar if TiO_2 NRs was composed by 3D ZnO (Figure. 5b). The exothermic peak at 350°C was shoulder and no exothermic peak at 700°C ascribed to rutile formation.

The samples were also characterized by X-ray powder diffraction (XRD) (Figure 6). Anatase phase TiO_2 as a single phase TiO_2 was obtained in the TiO_2-ZnO composite (1:1 of w/w ratio) upon heating the sample until 400°C. TiO_2 NRs have a crystal size 45.4nm, while crystal size of TiO_2 NRs–ZnO composite only gain a crystal size of 41,5nm. Those conclude that there is constrain crystal growth in the composite.

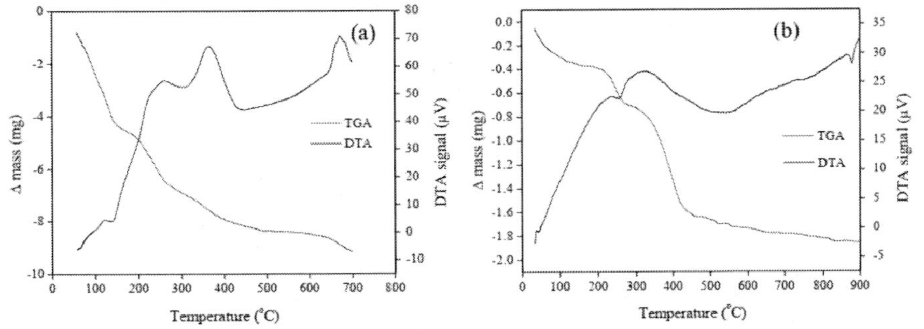

Figure 5. TG/DTA curves of (a) TiO_2 NRs (b) TiO_2 NR_2 - ZnO composite, recorded in air ambient.

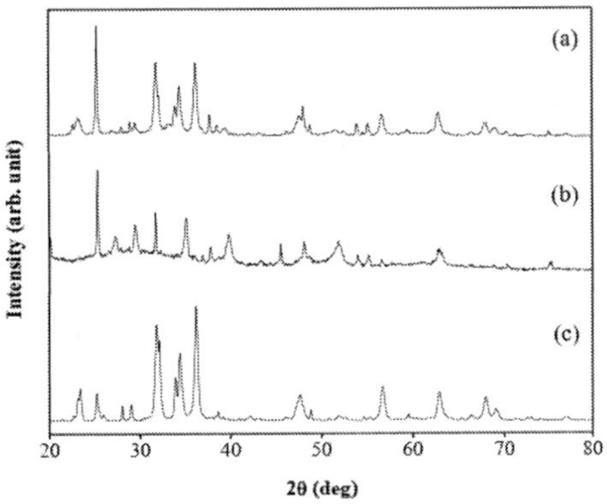

Figure 6. XRD patterns of (a) TiO_2NRs –ZnO composite, (b) TiO_2 NRs, and (c) ZnO NPs

3.2. *Performa TiO_2 nanorods - ZnO composite for DSSCs photoanode*

Figure 7 shows the current (I) - voltage (V) curves of the dye sensitized solar cells based on the TiO_2 NRs-ZnO composite with N3 dye. The values of short-circuit current density (I_{sc}), open-circuit voltage (V_{oc}), P_{max}, and overall power conversion efficiency (η) are summarized in Table 1.

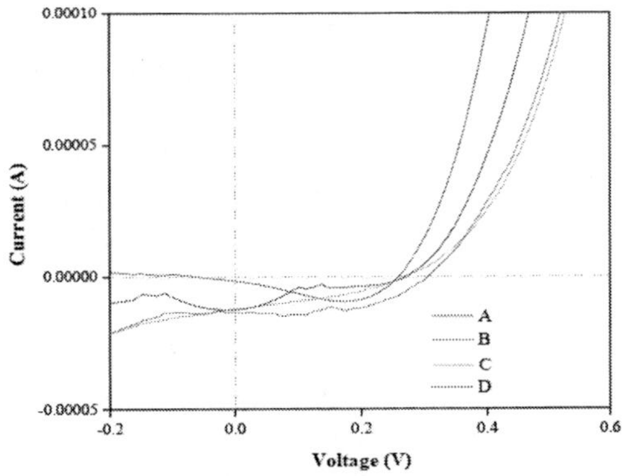

Figure 7. Current -voltage curves of the dye sensitized solar cells (DSSCs) based on (A) TiO_2 NRs photoanode, and TiO_2 NRs-ZnO composite photoanode (B) molar ratio of TiO_2 NRs:ZnO composite = 1:1, (C) molar ratio of TiO_2 NRs:ZnO composite = 1:2, (D) molar ratio of TiO_2 NRs:ZnO composite = 2:1.

Table 1. Performance of the solar cells with the electrode made of TiO_2 (A) TiO_2 NRs photoanode, and TiO_2 NRs-ZnO composite photoanode (B) molar ratio of TiO_2 NRs:ZnO composite = 1:1, (C) molar ratio of TiO_2 NRs:ZnO composite = 1:2, (D) molar ratio of TiO_2 NRs:ZnO composite = 2:1.

Sampel	Parameter					
	I_{sc} (μA)	V_{oc} (mV)	I_{max} (μA)	V_{max} (mV)	P_{max} (mW)	η (%)
A	0,785	0,092	3,57	200,01	0,715	0,36
B	1,28	0,313	11,8	200,07	2,37	1,18
C	1,15	0,045	6,66	175,84	1,17	0,59
D	2,04	0,164	9,12	0.187,53	1,71	0,86

It should be noted that the enhanced power conversion efficiency in TiO_2 NRs-ZnO composite solar cells could be attributed to (1) the suitable oriented TiO_2 nanorods and 3D ZnO for the hybrid solar cell, exciton diffusion length, and their compositions (2) the oriented TiO_2 nanorods and 3D ZnO reduced the electron recombination by provided a direct pathway for fast electron transport to the charge collecting electrode. In addition, the further improved the photoelectrochemical electrode (PCE) could be with the consideration of the ideal TiO_2 nanorods grown perpendicular to the ITO substrates via manipulating the densities of TiO_2 nanorods [13].

4. Conclusion

Material TiO_2 NRs has been synthesized from TiO_2 nanoparticle that has been treated ball milling, then has been further treated by hydrothermal refluxing method at a certain temperature and NaOH 12 M. Structural transformation to 1D nanorods capable increase surface area. Increasing of surface area may be increase dye adsorption onto TiO_2 surface and TiO_2 NRs have a crystal size of 45.4 nm, while that in TiO_2 NRs–ZnO composite only gain a crystal size of 41,5nm. Difference TiO_2 NRs-ZnO composite materials were investigated as a good photovoltaic materials. The enhanced power conversion efficiency in this TiO_2 NRs–ZnO composite solar cells could be attributed to the suitable oriented TiO_2 NRs–ZnO for the hybrid solar cell, exciton diffusion length, and the oriented TiO_2 nanorods reduced the electron recombination by provided a direct pathway for fast electron transport to the charge collecting electrode.

References

[1] Zhao L, Xia M, Liu Y, Zheng B, Jiang Q and Lian J 2012 *Mater. Trans.* **53** 463-468
[2] Uzunova-Buinova M, Dimitrove D, Radev D, Bojinova A and Todoravsky D 2008 *Mater. Chem. Phys.* **110** 291
[3] Lorret O, Francova D, Waldner G and Stelzer N 2009 *Appl. Cat.* B. **91** 39
[4] Donkova B, Dimitrov D, Kostadinov M, Mitkova E and Mehandjiev D 2010 *Mater. Chem. Phys.* **123** 563
[5] Jayasankar M, Hazem S, Salah H, Matthias D and Es-Souni M *Catal. Sci. Technol.* **2** 379-385
[6] Greene L E, Law M, Yuhas B D and Yang P 2007 *J. Phys. Chem. C.* **111** 50
[7] Wang M, Huang C, Cao Y, Yu Q, Deng Z, Liu Y, Huang Z, Huang J, Huang Q, Guo W and Liang J 2009 *Appl. Phys.* **42** 155
[8] Law M, Green L, Johnson J, Saykally R and Yang P 2005 *Nature. Mater* **4** 455-459
[9] Wahyuningsih S, FAdilah G, Hidayat R and Ramelah AH 2016 *Procedia Chemsitry* **19** 632-637
[10] Dutta H, Sahu P, Pardhan S K and De M 2002 *Mater. Chem. Phys.* **77** 153-164
[11] Rezaee M, Khoie S M M and Liu K H 2011 *Cryst. Eng. Comm.* **13** 5055-5061
[12] Tsai C C and Teng H 2006 *Chem. Mater* **18** 367-373
[13] Yodyingyong S, Zhou X, Zhang Q, Triampo D, Xi J, Park K, Limketkai B and Cao G 2010 *J. Phys. Chem. C.* **114** 21851–21855

Design of Hybrid Solar and Wind Energy Harvester for Fishing Boat

D A Banjarnahor[1], M Hanifan[1], E M Budi[1],
[1] Gedung TP Rachman Labtek VI Prodi Teknik Fisika, Jl. Ganesa no.10, Bandung 40132, Indonesia

E-mail: ditabanjarnahor@gmail.com

Abstract. In southern beach of West Java, Indonesia, there are many villagers live as fishermen. They use small boats for fishing, in one to three days. Therefore, they need a fish preservation system. Fortunately, the area has high potential of solar and wind energy. This paper presents the design of a hybrid solar and wind energy harvester to power a refrigerator in the fishing boat. The refrigerator should keep the fish in 2 – 4 °C. The energy needed is 720 Wh daily. In the area, the daily average wind velocity is 4.27 m/s and the sun irradiation is 672 W/m^2. The design combined two 100W solar panels and a 300W wind turbine. The testing showed that the solar panels can harvest 815 – 817 Wh of energy, while the wind turbine can harvest 43 – 62 Wh of energy daily. Therefore, the system can fulfil the energy requirement in fishing boat, although the solar panels were more dominant. To install the wind turbine on the fishing-boat, a computational design had been conducted. The boat hydrostatic dimension was measured to determine its stability condition. To reach a stable equilibrium condition, the wind turbine should be installed no more than 1.7 m of height.

1. Introduction

Rancabuaya is a small coastal village located in southern Garut, West Java. Based on the survey that had been conducted by us, the majority of people there live as fishermen. When fishing, they need electricity to light the boat and especially to store the fishes in refrigerator. To this day, fishermen there use fossil fuel, which contributes to air pollution, to fulfil the electricity needed. Meanwhile, Rancabuaya has a high potential of solar and wind energy. The daily average wind velocity is 4.27 m/s and the sun irradiation is 672 W/m^2. To meet the daily energy needed in a fishing boat and to utilize these renewable energy resources, a hybrid energy harvester is designed. This hybrid energy harvester will be installed on the fishing boat; thus, the fishermen can harvest and consume the energy while fishing. The system includes the energy storage system, energy monitoring system, and maximum power point tracking to enhance the efficiency of energy harnessed. The choosing of solar panel, battery, and charge controller depend on the energy produced by wind turbines used. Based on experiment that had been conducted [1], the amount of this energy is about 24 Wh on average.

---

[1] To whom any correspondence should be addressed

2. Experimental

2.1. Battery selection

The load, which is a cooling machine, needs an input voltage of 12 V and 180 W power. It will be turned on for four hours (4 h) a day, whenever fishes need to be cooled. If this cooling machine is needed at night, the energy would be supplied by battery only, since there is no solar irradiation, and the wind energy will only contribute a little. Considering this worst scenario, the battery capacity needed would be: (180 W)(4 h)/12 V = 60 Ah (Ampere hour). To keep the battery long-lasting, it should not be used too much. The ideal percentage of usage of a battery is 60%. Therefore, the battery capacity should be: 60 Ah/60% = 100 Ah. The 100 Ah capacity can be obtained by stringing three batteries parallel with capacity of 35 Ah each. This configuration is chosen because it will keep the voltage at 12 V, the same value as the cooling machine input voltage, and it minimizes the current losses.

2.2. Solar panel selection

The daily energy needed is 180 W x 4h = 720 Wh, where a power of 2 W is provided by the wind turbine [1]. The wind blows for 24 hours a day in Rancabuaya, so the wind turbine can operate 24 hours a day. That means, the energy from wind turbine equals to 2 W x 24 h = 48 Wh. Therefore, the energy needed from solar panel is 720 Wh – 48 Wh = 672 Wh. To fulfil that energy, the amount of power that solar panel should provide is 672 Wh/5 h = 134.4 W. Considering simplicity in installation, and also the losses (temperature, battery charging and discharging process, and losses in charge controller and wiring [2]), a 200 W output solar panel is chosen.

For this, four solar panels, 50 W each, can be installed. The second alternative is installing two solar panels, 100 W each. The third is choosing one solar panel with 200 W output. The second alternative is chosen because of the simplicity in installation and because of its lighter mass, regarding the boat stability. Besides, a four solar panel modules configuration will inflict more losses compared to two modules only. While a 200 W single module solar panel is the easiest to install, it is rarely sold in the market.

With only two modules, the solar panel can be stringed series or parallel. The parallel configuration gives a much safer system because one module is not affected by another, not like series configuration. For these reasons, a parallel configuration is chosen. The solar panel chosen is a monocrystalline solar cell type due to its high efficiency and its high number of availability in the market.

2.3. Charge controller

Charge controller chosen is hybrid charge controller for solar panel and wind turbine, Mars Rock type 750W50. Its function is to adjust current and voltage from both solar panel and wind turbine to battery specification, DC current with 12 V input. Charge controller also control the charging and discharging process to avoid overcharging and overdischarging.

From the charge controller specification, it is known that if the battery voltage reaches 16 V, the current from power supply will be disconnected so that the charging process stopped. While if battery reaches 10.8 V, the current flowing to the load will be disconnected so that the discharging process stopped.

This charge controller uses Pulse Width Modulation (PWM) type. It works whenever battery reaches overcharged point. In this condition, charge controller will turn the current connection on and off. The higher the battery voltage, the smaller the duty cycle is. When battery voltage reaches 16 V, the current flowing into the battery will be disconnected (duty cycle value = 0).

2.4. Design of data logger

Data logger is the instrument used to record, save, and display power and energy harnessed from solar panel and wind turbine. It uses current sensor ACS712 and voltage sensor 25 V as inputs. Since there are two points of measurement, the data logger will need two current sensors and two voltage sensors. The measurement data is processed with Arduino Uno to later be saved in SD Card and displayed using LCD 16x2. RTC DS1307 is used as time indicator in Arduino. Energy for this processing is provided by the battery, using a voltage regulator 12 V – 5 V. **Figure 1** below is schematic of the data logger designed.

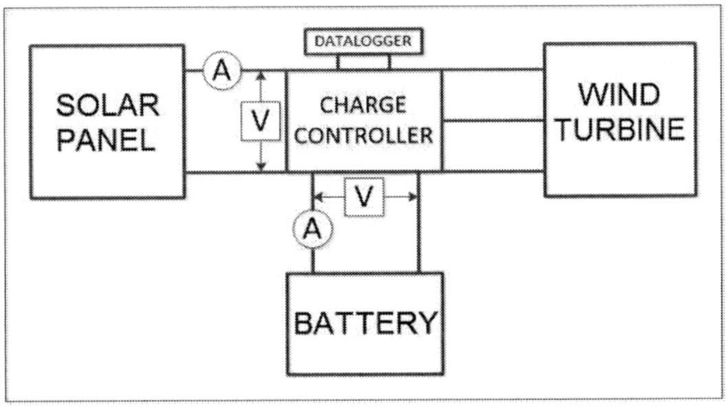

Figure 1. Data Logger Design Schematic.

2.5. Boat stability design

Figure 2 is boat design on Solidworks, of which the dimension and shape are based on real boats usually used in Rancabuaya beach.

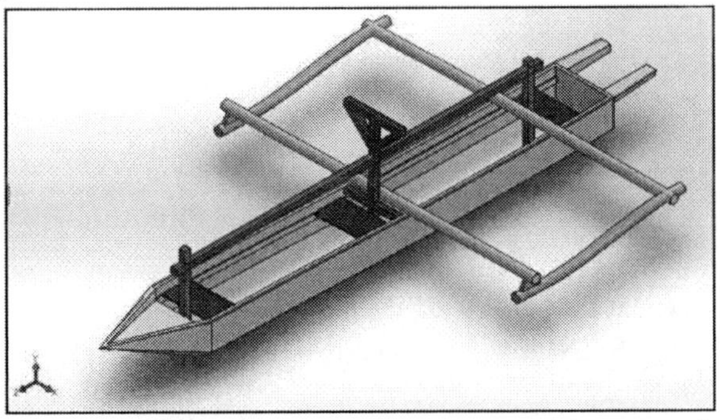

Figure 2. Boat Design on Solidworks Without Loads.

Boat stability condition is determined by position of mass centre (G), point of buoyancy (B), and point of metacentre (M) [3]. If M is above G, the boat will have enough force moment to restore the boat to **stable equilibrium condition**. If M is precisely coinciding with G, the boat will have zero force moment so that when given a disruption, the boat tilted and stay in that condition. This is known as **neutral equilibrium condition**. While if M is below G, the boat will have a negative force moment so that when given a disruption, the boat tilted and keep being tilted until it sinks. This is known as **unstable equilibrium condition**. This means that the condition we are trying to achieve is the stable equilibrium. In this case, the disruption given is adding loads to the boat.

After given the load consisting of two modules of solar panel, wind turbine, cooling machine, battery, and charge controller, the position of mass centre on boat changes. Using "mass properties" feature on Solidworks, the position of mass centre before and after the loading is obtained. To know whether the boat is sinking or not after loading, mass properties from each condition is observed. Mass property shows the mass and volume of boat and salt water. The mass, volume, and area surface is obtained from material type of boat's component and the dimension of the boat given. The material given is based on material used on real boat.

One parameter that has the biggest effect on stability is the height of wind turbine installation on boat. The higher the installation, the lower the stability is, and vice versa. Thus, the design sets the upper limit at 2 m and the lower limit at 1.7 m. The height above 2 m will give an unsteady equilibrium as will be shown in the simulation, while the height below 1.7 m will collide with middle pole, as shown in Figure 2.

3. Results and Discussion

3.1. Solar panel testing result

Solar panel testing in ITB area aims to obtain the relationship between output power and irradiance per time unit. The following is the data recorded using data logger, shown in **Figure 3**.

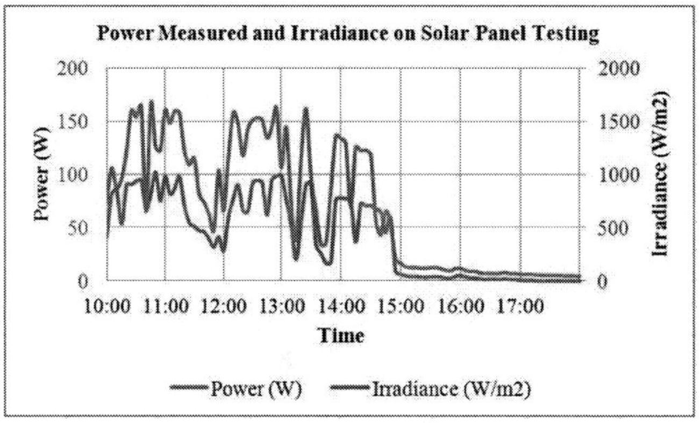

Figure 3. Solar Panel Power Measurement on August 18 2016.

Figure 3 shows measurement data on August 18 2016 from 10 a.m to 5 p.m. Since the relationship between power and irradiance shown is linier, the equation of solar panel power can be obtained, using first order regression, as shown below.

$$Solar\ panel\ power = 0.15\ x\ Irradiance + 9.03 \quad (1)$$

3.2. Integrated system testing result

System testing using Wind Turbine 2 (System 2) was done on August 27 2016, at ITB Research and Development rooftop, from 11.30 a.m to 2.50 p.m. While system testing using Turbine 1 (System 1) was conducted on August 30 2016. During testing, the sky condition varied a lot, from cloudy, overcast, to bright sky. From power measurement using data logger, the graph in **Figure 4** is obtained.

IOP Conf. Series: Earth and Environmental Science **75** (2017) 012007 doi:10.1088/1755-1315/75/1/012007

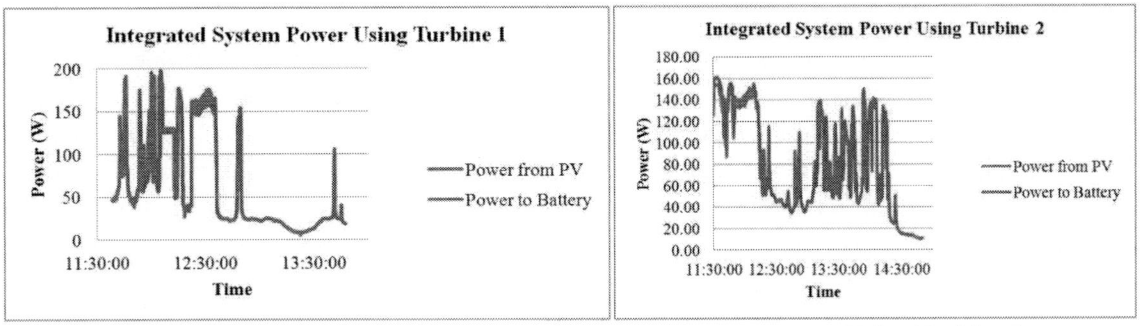

Figure 4. (a) System Power Using Turbine 1, **(b)** System Power Using Turbine 2

In System 1, the average power produced using solar panel is 60.17 W, while average power getting into the battery is 61.75 W. While in System 2, the average power produced by solar panel is 77.47 W and the power getting into the battery is 79.94 W. From the above experiments, and by using output power equation, the power output of the whole system in a day can be determined, as shown in **Figure 5** below.

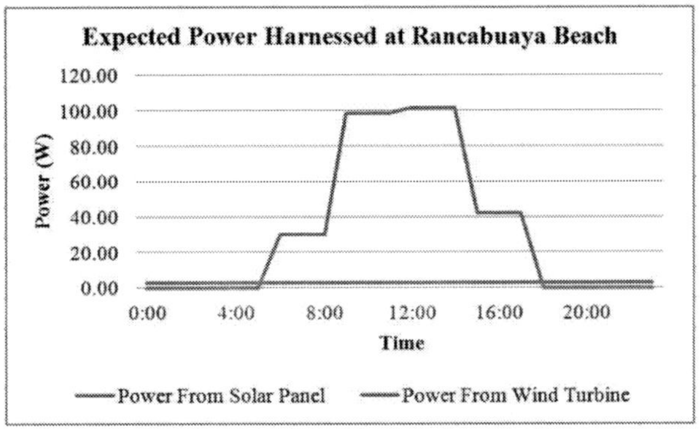

Figure 5. Rancabuaya Output Power Expected from The Integrated System.

Using Turbine 1, the total energy harvested is 858 Wh, with 95% harnessed by solar panel and 5% from wind turbine. While using Turbine 2, the total energy harvested is 877.92 Wh, with 93% comes from solar panel and 7% from wind turbine.

3.3. Boat stability simulation result

The followings are G, B, and M position before and after the loading.

3.3.1. 2 Meter height of wind turbine installation

Table 1 shows mass properties of **Boat** and **Salt Water** after the loading, with the height of wind turbine installation: 2 m.

Table 1. Boat Floatation Position Toward Salt Water.

Boat Mass Properties	Salt Water Mass Properties
Mass = 1498.94 kilograms	Mass = 1995.52 kilograms
Volume = 1.95 cubic meters	Volume = 1.95 cubic meters
Surface area = 68.54 square meters	Surface area = 68.54 square meters

For a same volume, salt water has a higher mass than the boat. This means that the boat has a lower density so that it can stay floating. The mass difference between them is 496.58 kg, which means, the boat can carry up to eight passengers with total mass 496.58 kg.

Next, the boat stability condition is determined by examining the position of G, B, and M before and after the loading. **Figure 6** and **Figure 7** are the illustrations. By looking at Figure 7, G coincides with M, meaning that the boat has a neutral equilibrium condition. Thus, a 2 m height of wind turbine installation would disrupt the boat stability.

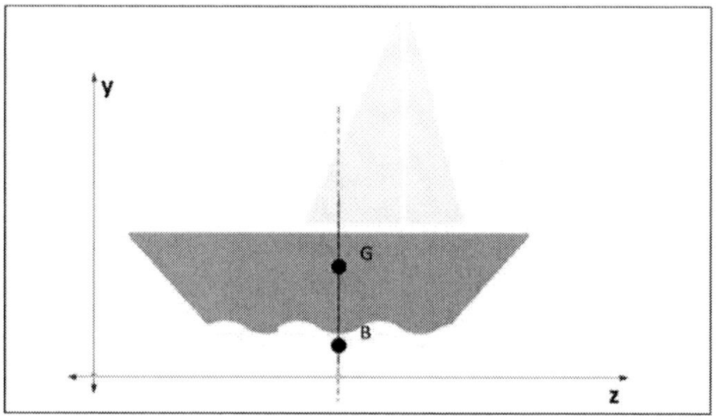

Figure 6. Boat Stability Condition Before The Loading.

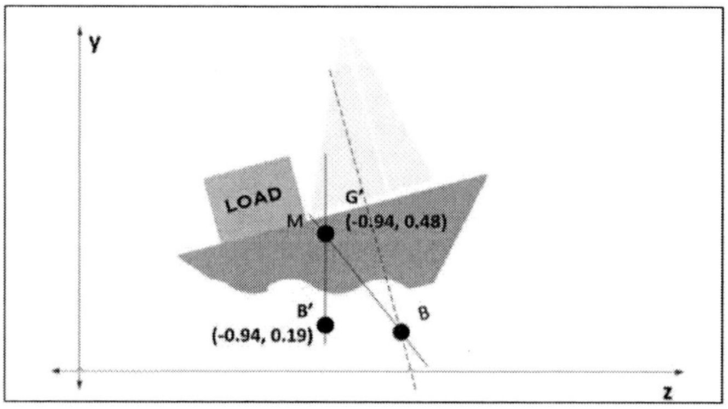

Figure 7. Boat Stability Condition After The Loading.

3.3.2. 1.7 Meter height of wind turbine installation

Table 2. Boat Floatation Position Toward Salt Water.

Boat Mass Properties	Salt Water Mass Properties
Mass = 1491.40 kilograms	Mass = 1994.53 kilograms
Volume = 1.95 cubic meters	Volume = 1.95 cubic meters
Surface area = 68.44 square meters	Surface area = 68.44 square meters

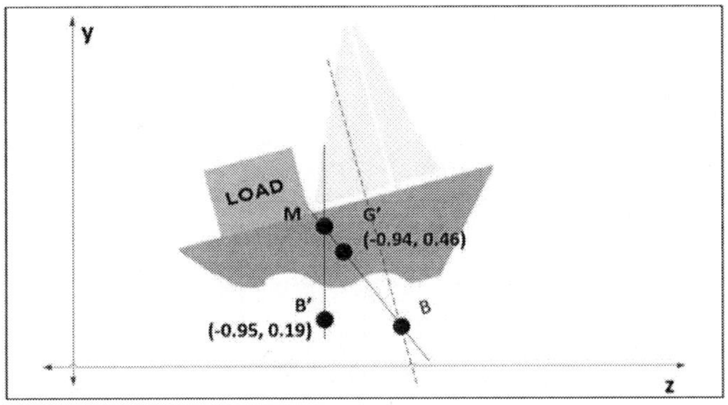

Figure 8. Boat Stability Condition After The Loading.

The M position that is above the G in **Figure 8** shows that the boat has enough amount of force moment to keep it stable when given a disruption. Thus, the boat has a stable equilibrium condition.

4. Conclusion

An energy harvester system has been designed to be installed on Rancabuaya fishing boat. It is expected that the total energy harnessed in a day is 858 Wh using Turbine 1, and 877.92 Wh using Turbine 2. The energy needed in Rancabuaya fishing boat itself is 720 Wh per day. Thus, the energy harvester system designed has been able to provide the energy needed. To store this energy, three batteries of which each capacity is 35 Ah are chosen. These three batteries are stringed parallel.
The boat will have a stable equilibrium condition after the loading if the wind turbine is installed on a 1.7 m height.

References

[1] Banjarnahor D, Hanifan M, Budi E 2016 *Perancangan Sistem Pemanen Energi Hibrid Angin dan Matahari pada Perahu Nelayan Rancabuaya* (Bandung: Institut Teknologi Bandung)
[2] Frank K and Gosmawi D 2015 *Handbook of Energy Efficiency and Renewable Energy* (CRC Press)
[3] Barrass B and Derrett D 2006 *Ship Stability* (Elsevier)

IOP Conf. Series: Earth and Environmental Science 75 (2017) 012008 doi:10.1088/1755-1315/75/1/012008

Study of Vertical Axis Wind Turbine for Energy Harvester in A Fishing Boat

E M Budi[1], D A Banjarnahor[1], M Hanifan[1]
[1] Gedung TP Rachman Labtek VI Prodi Teknik Fisika, Jl. Ganesa no.10, Bandung 40132, Indonesia

E-mail: mursito@tf.itb.ac.id

Abstract. The wind speed in the southern beach of West Java Indonesia is quite promising for wind energy harvesting. A field survey reported that the wind speed reached 10 m/s, while the average recorded in a year is about 4.7 m/s. In this study, two vertical axis wind turbines (VAWT) were compared to be used in that area through calculation as well as experiments. The experiments measured that the turbines can produce about 7.82W and 2.33W of electricity respectively. These experiments are compared with theoretical calculation to obtain the performance of both turbines used. The coefficient of performance (cp) experimentally is 0.09 for Turbine 1 (hybrid Savonius-Darrieus rotor) and 0.14 for Turbine 2 (Savonius rotor). While, rotor's mechanical performance Cpr, obtained theoritically through calculation, is 0.36 for Turbine 1 and 0.12 for Turbine 2. These results are analysed from mechanical and electrical view.

1. Introduction

Rancabuaya village, Garut, West Java, is a village located in the coastal area. Based on the survey that we had conducted, the majority of people there work as fishermen. During the sail, fishermen need electricity to support their activities. For example, for lighting and storing the fishes in refrigerator. Nowadays, fishermen using diesel fuel, which is a contributor to air pollution. Hence, the supply of electrical energy using renewable resources is a solution that can be offered. Solar panels and wind turbines will be used to harvest energy. Rancabuaya beach itself has a high potential of solar and wind energy. The daily average wind velocity is 4.27 m/s and the sun irradiation is 672 W/m^2 [1]. This paper focused on the study of wind turbine performance on the boat, which will be reviewed in September, as the fishing season is in September. The wind turbines used are Vertical Axis Wind Turbine. Turbine 1 uses hybrid Savonius-Darrieus rotor, while Turbine 2 uses Savonius rotor. Both will be compared aerodynamically and electronically, based on the electricity produced on application and on theory.

2. Experimental

2.1. Wind turbine testing

[1] To whom the correspondence should be addressed

Wind turbines brand Mars Rock with models VAWT-VA300W1224-A type hybrid Savonius-Darrieus would later be called as **Turbine 1** and model VAWT-VA300W1224-S type helix Savonius would later be called as **Turbine 2**. Both turbines used has a maximum power of 300 W and can start to rotate at a wind speed of 1.3 m/s. Maximum power is obtained at an average wind speed of 10 m/s.

2.1.1. Mechanical power
Mechanical power carried by the wind at a certain speed is expressed by the following equation [2].

$$P = \frac{1}{2}\rho A v^3 \tag{1}$$

At a height of 50 m, data that was collected by NASA satellite [1] shows the average wind speed in September on the Rancabuaya beach is 5.91 m/s. By using the speed comparative law by altitude [3]:

$$\frac{U(z)}{U(z_r)} = \left(\frac{z}{z_r}\right)^{\alpha} \tag{2}$$

$U(z)$: wind speed (m/s)
$U(z_r)$: reference wind speed (m/s)
z : height (m)
z_r : reference height (m)
α : surface roughness
Where:

$$\alpha = \frac{1}{ln\frac{z}{z_0}} \tag{3}$$

z_0 : surface roughness parameter
Z_0 value depends on the roughness of the surface where wind turbines are installed. For water surface, z_0 value is 0.0001.

Hence, the wind speed at a height of 2 m (target altitude of wind turbine installation on the boat) can be calculated. By comparing the speeds at a height of 50 m, average wind speed obtained at a height of 2 m is 4.27 m/s. On experiment using wind tunnel, the value that approaches the speed of 4.27 m/s is 4.6 m/s. Values of air density ρ depends on the air temperature at the time. The average air temperature at Rancabuaya beach at a height of 2 m is 24.3°C. By interpolation of the air density data at a temperature of 20°C and 25°C, obtained $\rho = 1.22$ kg/m^3. Using equation (1), the comparison of mechanical power between Turbines 1 and 2 is shown in **Table 1**.

Table 1. Mechanical Power by Wind for Both Turbines.

	Turbine 1 (W)	Turbine 2 (W)
Area = A (m^2) (Approached as triangle, a = 1 m, and t = 0.75 m)	A_1 = 0.375	A_2 = (0.75)(0.4) = 0.3
Mechanical Power = P (W)	22.27	17.81

2.1.2. Electrical power measured
Voltage and current measurement results are obtained from the sensors that are connected to the Arduino microcontroller. In this calculation, the measured power is calculated by the equation $P = VI$, in units of Watts, as shown in the **Table 2**.

Table 2. Electrical Power Measured from both Turbines.

	Turbine 1 (W)	Turbine 2 (W)
Wind Speed (m/s)	4.6	4.6
Voltage (V)	4.34	7.00
Current (A)	0.48	0.40
Power (W)	2.06	2.83

Figure 1. Wind Turbine Testing Using Wind Tunnel.

3. Results and Discussion

The tests aim to determine the performance of each wind turbine for each wind speed. This performance is shown by two coefficients: Cp (coefficient of performance) and Cpr (coefficient of rotor performance).

$$c_p = P_{measured}/P_{max} \tag{4}$$

$$c_{pR} = P_{rotor}/P_{max} \tag{5}$$

The graphs in Figure 2-7 are the test results that demonstrate the performance of each wind turbine, compared to TSR. TSR stands for Tip Speed Ratio, and is a comparison of angular velocity of rotor to wind velocity. TSR value will determine whether a wind turbine has reached a self-start condition or not. Self-start is a condition where the rotor has accelerated to a steady condition, where it produces significant amount of power [4]. Self-start happens when TSR > 1.

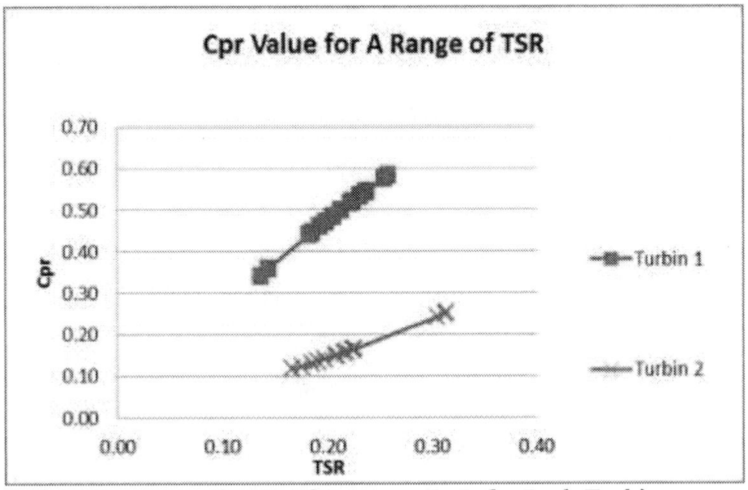

Figure 2. Relation of Cpr over TSR for Both Turbine.

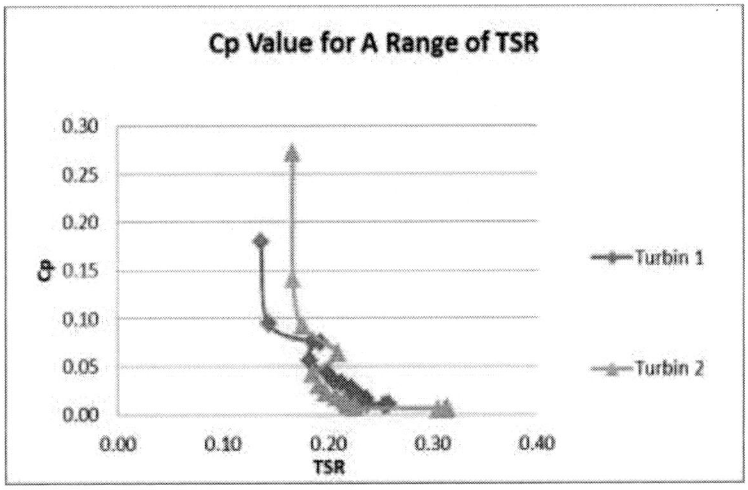

Figure 3. Relation of Cp over TSR for Both Turbine.

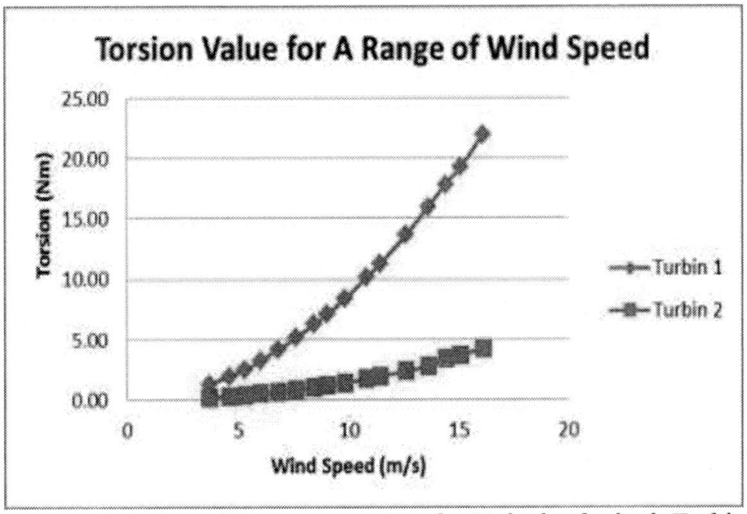

Figure 4. Relation of Torque over Angular Velocity for both Turbine.

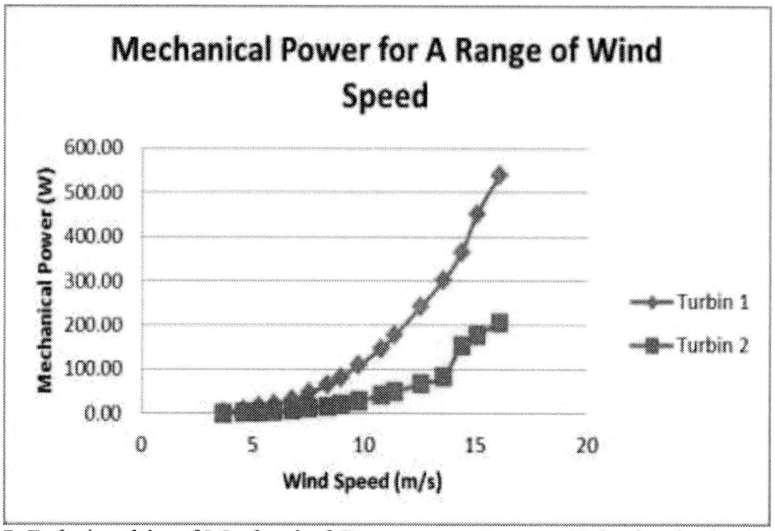

Figure 5. Relationship of Mechanical Power over Anguar Velocity for both Turbine.

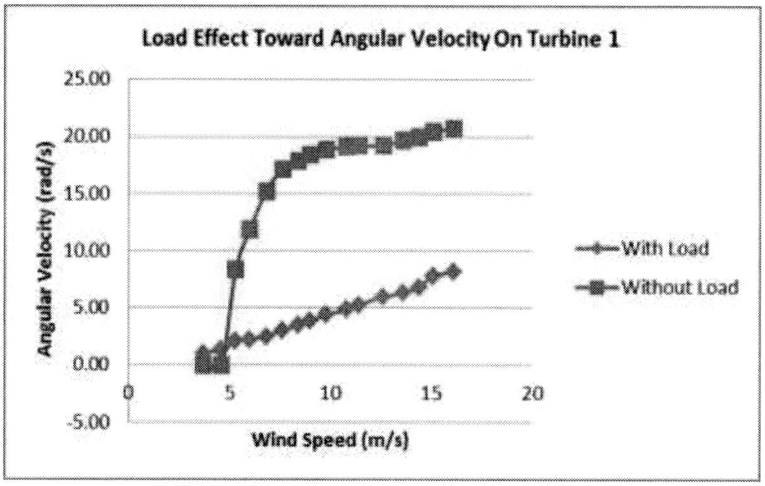

Figure 6. Effect of Loading to Angular Velocity for Turbine 1.

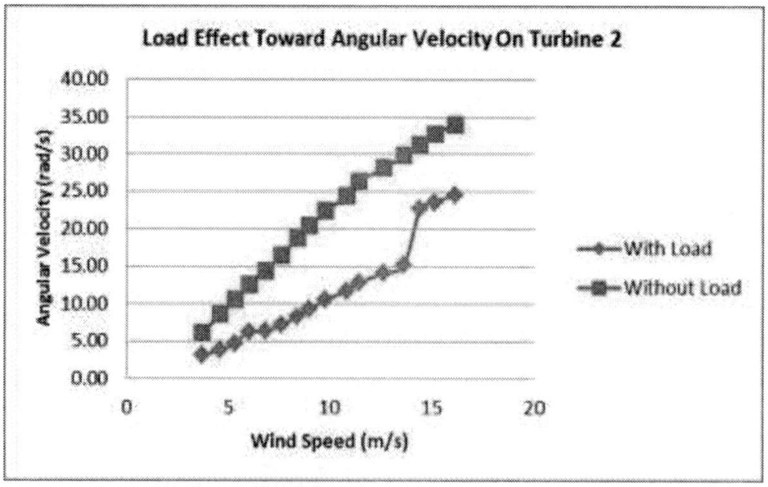

Figure 7. Effect of Loading to Angular Velocity for Turbine 2.

The output power obtained is only ± 2W at wind speeds of 4.6 m/s for each wind turbine. According to the calculations, the range of TSR on Turbine 1 is only on the range of values 0.14 – 0.25 and on the value of 0.17 – 0.31 for Turbine 2. We can conclude that both turbines fail to reach the self-start condition. This failure will be analyzed from the view of aerodynamics of rotor and effect of electrical loading.

3.1. Wind turbine aerodynamics performance

To understand the aerodynamic conditions of rotor as spinning objects, the force, direction, and speed that works on it need to be reviewed. Figure 8 depicts these parameters.

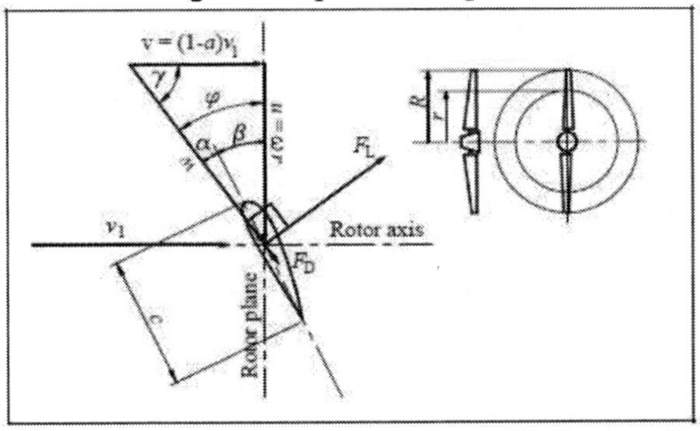

Figure 8. Vector of Direction, Speed, and Force Worked on One Blade [5].

Based on Figure 8, the force that drives the rotor to keep spinning is FL, the lift force. Lift Force themselves is perpendicular to the drag force FD. Both of these force contribute to the magnitude of the resultant torque, as described in **Figure 9**.

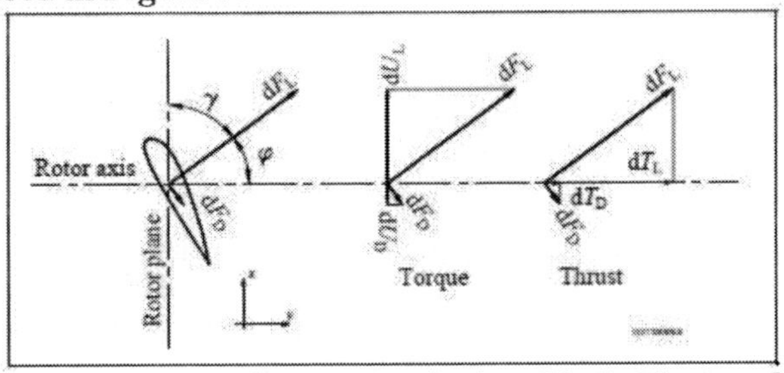

Figure 9. Vector of Force Worked on Blade Element [5].

On the x-axis and y-axis, which is the rotor axis and rotor field, there are two vector forces work. From these two forces, the one contributes to generate torque is dU. Where [5]:

$$T = F\,x\,r \tag{6}$$

T : torsion (Nm)
F : tangential force (N)
r : force arm (m)

Because of the lift force and drag force, the force components dU are also divided into dUd and dUl. Then the resultant force becomes dU. Thus, the components of this force will become maximum when the lift force is maximum and drag force is minimum. Value of dU will be integrated, with each element gives the overall force U. Maximum force U will provide the maximum torque value to produce maximum mechanical rotor power. Mechanical rotor power itself is given by the equation

$$P_{rotor} = T\omega, \tag{7}$$

where ω is the angular speed of the rotor. If the rotor consist of B blades, the power equation becomes

$$P_{rotor} = BT\omega. \tag{8}$$

Thus, to maximize the rotor power, parameters that need to be reviewed is the torque and angular speed.

Based on the data shown in Figure 2 and 3, it is found that Cp value of Turbine 2 is greater than the Turbine 1. While, the Cpr value of Turbine 1 is greater than 2. This means, aerodynamically, (in theory) Turbine 1 shows a better performance than the turbine 2. This is supported by the fact that Turbine 1 has more blades and larger cross-sectional area, so the wind power harvested is higher. Meanwhile, Cp comparison shows that the performance of Turbine 2 is better than the Turbine 1. This is because both turbines failed to reach self-start.Tthe maximum power that can be obtained by Turbine 1 in theory is higher than Turbine 2 because of larger cross-sectional area; while in practice, both turbine produce the same amount of power. Hence, Cp of Turbine 1 is lower than Turbine 2.

Based on Figure 4 and 5, it can be seen that Turbine 1 produce torque and rotor power that are higher than Turbines 2 at the same wind speed. It means that Turbine 1 has resultant force U larger than Turbine 2. This is caused by the design and the number of blades of each turbine.

Turbine 1, Darrieus-type blades, generate lift force greater than Turbine 2, Savonius type. Darrieus-blades type are generally designed to have a certain angle which gives a bigger lift force so that the maximum torque is greater than the Savonius-blades type.

Turbine 2 generate lift force smaller (or smaller resultant force U), because the shape of the blade is like a spoon. When the concave side faces the direction of the wind, the blades provide a drag force that blocks the rotor rotation. At the same time, the convex side that is also facing the direction of the wind generate lift force that push the rotor to keep rotating. Because these two forces work together, the resultant force of the two forces is small, so that the torque and mechanical power generated is also small. It is also shown by the value of Cpr of Turbine 2 which is much smaller than the first turbine, especially at higher speeds.

3.2. Effect of electrical loading

As already mentioned, the low output power produced by both turbine is because of failure to reach the self-start. This failure can also be caused by the load experienced by the rotor's generator and the electrical load that was connected to the rotor. In this case, the load of the electricity used is 12 V 35 Ah battery, charge controller, and 180 W refrigerator. These loads give significant effect on inhibiting the rotor rotation, as shown in Figure 6 and Figure 7.

It is found that the angular velocity ratio of Turbine 1 and Turbine 2 are not in accordance with the theory of the rotor. Turbine 1 should give a higher speed than the Turbine 2, because theoretically it gives a higher lift force. This can be caused by several things. Turbine 1, which is a hybrid type failed to achieve self-start even with the help of Savonius type rotor. In fact, this Savonius rotor may also act as a load to the rotation of the Turbine 1's rotor, because two rotors are connected by the same hub to the generator. Because the self-start condition has not been reached, the lift force generated by Darrieus rotor can not balance the loads that it receives and cause the rotor angular velocity lower.

Other causes which give effect to the failure of the self-start is the generator that does not function as it should. In other words, it has low level of efficiency.

4. Conclusion

Hybrid type wind turbine, Darrieus-Savonius (Turbine 1) and Savonius type (turbine 2) used in this study showed a low performance. Indicated by the overall performance Cp which is 0.09 for Turbine 1 and 0.14 for Turbine 2. While, rotor's mechanical performance Cpr is 0.36 for Turbine 1 and 0.12 for Turbine 2. The value Cp and Cpr is at wind speeds of 4.6 m/s. For higher wind speeds, it is obtained that the both coefficient value had declined.

Based on these values, in theory, Turbine 1 produced mechanical power greater than turbine 2. Meanwhile, if it observed from its overall performance, Turbine 2 showed a better performance.

References

[1] Banjarnahor D, Hanifan M, Budi E 2016 *Perancangan Sistem Pemanen Energi Hibrid Angin dan Matahari pada Perahu Nelayan Rancabuaya* (Bandung: Institut Teknologi Bandung)
[2] Hau E 2005 *Wind Turbines* (Berlin: Springer)
[3] Benoit R 1977 *J. appl. Met* **16** 859-860
[4] Hill N, Dominy R, Ingram G, Dominy J 2009 *Proc. IMechE Part A: J. Power and Energy* **223** 21-29
[5] Gundtoft S 2009 *Wind Turbines* (Aarhus: University College of Aarhus)

Preparation of Activated Carbon from Palm Shells Using KOH and $ZnCl_2$ as the Activating Agent

Yuliusman[1], Nasruddin[2], M K Afdhol[1], R A Amiliana[1], A Hanafi[1]
[1]Teknik Kimia, Universitas Indonesia, Kampus Baru UI Depok, Indonesia
[2]Teknik Mesin, Universitas Indonesia, Kampus Baru UI Depok, Indonesia

E-mail: usman@che.ui.ac.id

Abstract. Palm shell is a potential source of raw materials for the produce of activated carbon as biosorbent for quite large numbers. The purpose of this study is to produce activated carbon qualified Indonesian Industrial Standard (SNI), which will be used as biosorbent to purify the impurities in the off gas petroleum refinery products. Stages of manufacture of activated carbon include carbonization, activation of chemistry and physics. Carbonization of activated carbon is done at a temperature of 400°C followed by chemical activation with active agent KOH and $ZnCl_2$. Then the physical activation is done by flowing N_2 gas for 1 hour at 850°C and followed by gas flow through the CO_2 for 1 hour at 850°C. Research results indicate that activation of the active agent KOH produce activated carbon is better than using the active agent $ZnCl_2$. The use of KOH as an active agent to produce activated carbon with a water content of 13.6%, ash content of 9.4%, iodine number of 884 mg/g and a surface area of 1115 m^2/g. While the use of $ZnCl_2$ as the active agent to produce activated carbon with a water content of 14.5%, total ash content of 9.0%, iodine number 648 mg/g and a surface area of 743 m^2/g.

1. Introduction

One of the purification method of the byproducts of petroleum refinery yields off gases is by adsorbing the impurities using activated carbon as a biosorbent [1]. Activated carbon is a biosorbent with high adsorption capabilities, carbonization-processed and chemically or physically activated. According to the structure pattern, activated carbon is an amorphous carbon consisted mostly of free carbons and has inner surfaces; resulting the high adsorption capabilities compared to other adsorbents [2]. Organic as well as inorganic compounds containing lignin, hemicelluloses, and cellulose can be utilized as the raw material of producing activated carbon because of the high effectivity for adsorption process. One of the highly potential and currently developed raw material for making activated carbon are palm shell.

Palm shell contains 26.6% cellulose and 27.7% hemicelluloses, which are good to be utilized as activated carbon. Palm shell is one of the largest waste produced from palm oil refinery, achieving 12% weight of palm fruit. The average production of palm fruit per year is 5.6 million tones, meaning there could be around 672,000 tonnes of shells produced. With the availability of the waste, an advanced process is needed to convert the palm shell waste into a high economical valued product [3].

The quality of activated carbon that can be converted into biosorbent on pollutant adsorption is strongly affected by the production condition. Parameters affecting activated carbon production process are dehydration, carbonization, and activation. On the other hand, adsorption process affectivity is strongly affected by several parameters including adsorption properties, temperature, contact time, and surface area. Previously, there has been a lot of method producing activated carbon from organic materials but the carbon produced has yet achieved the required standard. Hence, this research aims on producing an activated carbon that has a good characteristic as a biosorbent.

Activation using activating agent on activated carbon production has been done by adding several chemicals like NaOH, H_3PO_4, dan $ZnCl_2$. The usage of H_3PO_4 produced activated carbon with the surface area of 438.9 m^2/g [4]. The usage of KOH as an activating agent can produce activated carbon with the surface area of 3000 m^2/g [5]. On the other hand, usage of $ZnCl_2$ as an activating agent can produce activated carbon with the surface area of 1100 m^2/g [6]. Activating agents KOH and $ZnCl_2$ are activating agents that can potentially produce activated carbon with a high surface area. Therefore, this research utilized activating agents KOH and $ZnCl_2$ in the production of activated carbon from palm shells.

2. Experimental

2.1. *Carbonization*

In the research, palm shells are used as the raw material of activated carbon. The research starts by doing a carbonization on the temperature of 400°C for 2 hours. The outcome of carbonization is palm shell charcoal that has been deprived of water content and volatile matters. Palm shell charcoal then shattered using mortar. The process continues to sieving with mesh to yield smaller sized particles (1-2 mm). Smaller particle size increases the charcoal's surface area, expected to increase the number of activated pores.

2.2. *Chemical and physical activation*

Carbons with the size of 1-2 mm is then chemically activated with activating agents KOH (75% KOH solution) and (25% $ZnCl_2$ solution). The mixing composition of activating agent and carbon is 1:4 (carbon : activating agent). Activation process is done with the stirring speed of 100 rpm, temperature ± 85°C for 2 hours. Activated carbon from the chemical activation then physically activated by using N_2 gas on 850 °C for 1 hour. Activation is done by staged heating in order to reach the desired activation temperature. N_2 gas is channeled with the flow rate of 100 cc/min. After 1 hour, physical activation is continued with channeling CO_2 gas for 1 hour on the temperature of 850°C [8]. After the activation process is finished, the activated carbon is cooled with channeling CO_2 gas into the reactor. The activated carbon yielded is then washed with 0,1 N HCN solution until it reaches neutral pH. The activated carbon is then washed with distilled water to get rid of the chloride. The size of the activated carbon is then reduced into 212-150 µm, 150-106 µm, 106-50 µm, 50-37 µm.

Dried activated carbon is characterized based on Indonesian Industrial Standard. Characterization includes activated carbon yield gain, iod number test, water content test, ash content test and losses from heating. To change the iodine number into surface area a linear regression equation is used that refers to ASTM D-4607-94 [7]. The equation is as follows

$$\text{Iod Number} = 0.6366 \times \text{surface area} + 174.34 \quad (1)$$

3. Results and Discussion

3.1. *Carbonization of palm shell*

Carbonization process occur on 400°C for 2 hours. The results of palm shell carbonization is displayed as yield percentage data on Table 1. From the table we can conclude that the process produced charcoal with the average yield percentage of 38%. Charcoal yield shows that volatile matters contained on the palm shell is quite high, causing large mass loss through the carbonization process. Charcoal yield produced is also an indication of the number of fixed carbon contained on palm shell.

Table 1. Palm shell carbonization yield

No	Initial Mass (g)	Final Mass (g)	Yield (%)
1	587	218	37.2
2	625	253	40.5
3	502	181	36.2
	Average		38.0

3.2. Chemical activation with activating agent KOH and physical activation

Choice of KOH as activating agent is based on the ability to open the pores of the carbon. Pores formation on the chemical activation is then completed on the physical activation through heating. The more pores formed, the more surface area produced on the activated carbon. Physical activation is a heating process on high temperature without the presence of free oxygen, by channeling N_2 gas into the reactor throughout the activation process, continued by channeling CO_2 gas functioning as cooler gas. Yield percentage gained is presented on Table 2. The average yield percentage is 94.3%, showing the number of activated carbon formed from physical activation process. The high activated carbon content is produced because of a carbonization process prior activation; removing volatile matters and isolating the material from air, thus producing high yield percentage [9].

Table 2. Physical activation process activated carbon yield (KOH)

No	Initial Mass (g)	Final Mass (g)	Yield (%)
1	100	94.3	94.0
2	101	95	94.2
3	101	96.1	94.5
4	101	95.7	94.7
	Average		94.4

3.2.1. Water content and ash aontent

Water content percentage of the activated carbon is 13.6%. According to Indonesian Industrial Standard, the minimum water content of activated is 15%. Therefore, the activated carbon produced from palm shell in the research has met the Indonesian Industrial Standard's criteria of water content. Ash content percentage of activated carbon on the research is 9.4%, which has met the Indonesian Industrial Standard's criteria of ash content of 10%.

3.2.2. Surface area using Iod number method

Activated carbon quality can be observed from the adsorption ability towards iodine. Activated carbon's adsorption ability towards iodine solution indicates its ability to adsorb components with low molecular weight. Activated carbon with high iodine adsorption ability has larger surface area and has larger micro- and mesoporous structure. There is a correlation between iod number and specific surface area of activated carbon, which can be determined using linear regression equation referencing on ASTM-4607-94. Figure1shows specific surface area based on activation treatment.

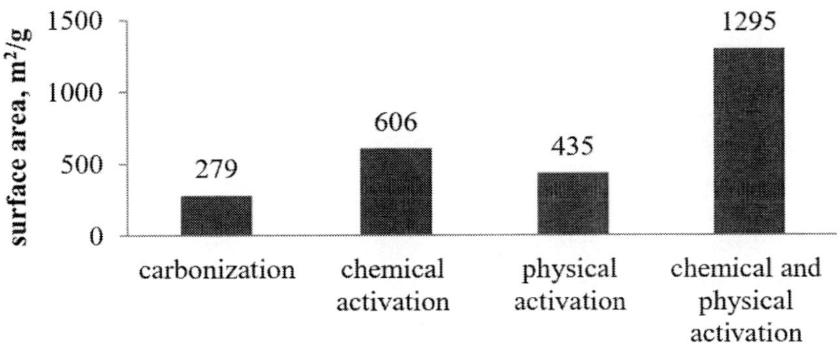

Figure 1. Effect of treatment on specific surface area (activating agent KOH)

From Figure 1, it can be seen that activation treatment increases specific surface area. Charcoals without activation has the smallest specific surface area. The principle of carbonization is to form pores, but the number of pores formed is still small compared to activation process. Small number of pores causes charcoal to be unable of adsorbing iodine solution effectively. The number of iod adsorbed is 352 mg/g and the surface area of the activated carbon is also small, which is 279 m^2/g. Activation process greatly affects the formation of surface area of activated carbon, the more complex the activation process will form more pores so that the activated carbon can absorb more iodine and the surface area formed will be greater. On the other hand, the effect of activated carbon particle size on surface area can be seen at Figure 2.

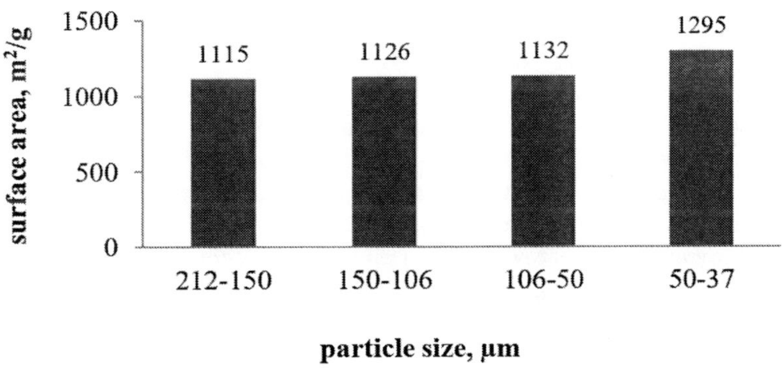

Figure 2. Effect of activated carbon particle size on specific surface area (activating agent KOH)

The difference in activation treatment on activated carbon sample affects the specific surface area significantly. But on the variation of the particle size, the specific surface area formed is not so different. This is caused by the formation of pores from the physical and chemical activation process, thus the variation of the pore quantity is uniform. There is a little difference from the specific surface area formed, because the smaller the particle size thus the number of particle will be greater and the contact with iodine solution will be more intense. The more particles contacted with iodine solution will cause an increase on iod number and surface area. But on the fine particles, the contact between activated carbon and iodine solution is not going smoothly because the carbon is on the surface of iodine solution.

3.3. Chemical activation with activating agent $ZnCl_2$ and physical activation

After size reduction, the palm shell will next be activated chemically by adding activating agent solution, $ZnCl_2$. Choice of $ZnCl_2$ is because it is acidic thus has the ability of opening pores on carbon by removing impurities on carbon. The process after wards is carbon activation by heating on 850°C with the absence of free oxygen by channeling CO_2 and N_2 gas into the reactor through out the activation process. Yield percentage gained is presented on Table 3. Activation process cause mass loss. The quantity of mass loss percentage is a representation of water content and impurities contained on the solid mixture between activating material/raw material and volatile matters that are still present, thus can be removed throughout the heating process.

Table 3. Physical activation process activated carbon yield ($ZnCl_2$)

No	Initial Mass (g)	Final Mass (g)	Yield (%)
1	100	78.4	78.4
2	100	76.7	76.7
3	100	77.9	77.9
4	100	78.1	78.1
	Average		77.1

3.3.1. Water content and ash content

Water content percentage of the activated carbon is 14.5%. According to Indonesian Industrial Standard, the minimum water content of activated is 15%. Therefore, the activated carbon produced from palm shell in the research has met the Indonesian Industrial Standard's criteria of water content. Ash content percentage of activated carbon on the research is 9.0%, which has met the Indonesian Industrial Standard's criteria of ash content of 10%.

3.3.2. Surface area using Iod number method

Activated carbon quality can be observed from the adsorption ability towards iodine. Activated carbon's adsorption ability towards iodine solution indicates its ability to adsorb components with low molecular weight. Activated carbon with high iodine adsorption ability has larger surface area and has larger micro- and mesoporous structure. Figure 3 shows the effect of treatment on specific surface area. Highest activated carbon surface area is gained when chemical and physical activation is done, which is 743 m^2/g. Meanwhile, the lowest surface area is gained on carbon without activation, which is 275 m^2/g.

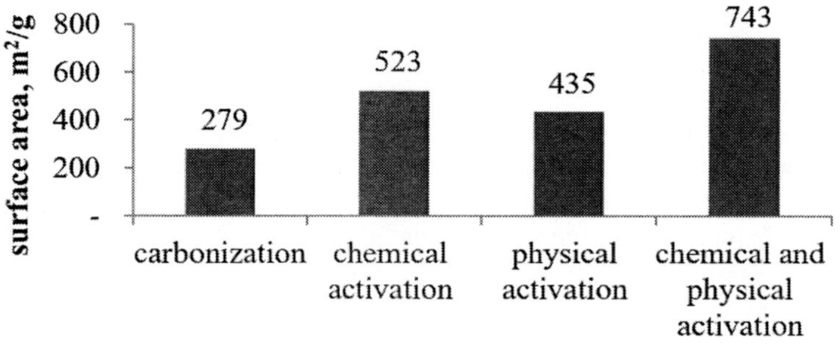

Figure 3. Effect of treatment on specific surface area (activating agent $ZnCl_2$)

Figure 4 shows the effect of particle size on activated carbon's surface area. Highest surface area is yielded on activated carbon with the particle size of 50-37 μm, which is 743 m^2/g. Particle size affects the activated carbon's surface area by a little. The smaller the particle size, the larger the surface area. Therefore, from the results gained we can get that the most optimal particle size of activated carbon in adsorbing iodine is 50-37 μm.

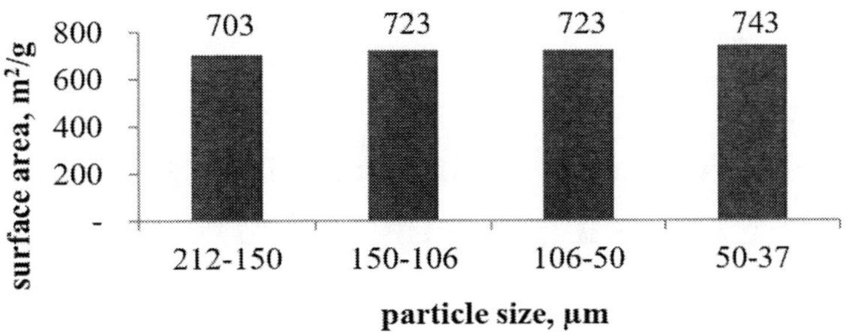

Figure 4. Effect of activated carbon particle size on specific surface area (activating agent $ZnCl_2$)

Research results show that activated carbon using KOH and $ZnCl_2$ can increase iodine adsorption ability (showing surface area). There is a high iodine adsorption ability increase between carbon without and activated carbon using KOH and $ZnCl_2$. This shows that in the activation process, the activating agent forms new pores so that the adsorption ability increases. Amorphous carbons prevent pores from reacting through early oxidation stage, resulting the opening of closed pores and forming new pores. Research results show that KOH is more effective on increasing the surface area than $ZnCl_2$ [10], shown by the surface area produced on activated carbon.

4. Conclusion

From the research results, data processing, and analysis; we can conclude that an activated carbon produced from the research has met the Indonesian Industrial Standard with carbon based on palm shell has a carbon yield of 38%. Chemical activation with activating agent KOH and physical activation on 850°C yields activated carbon with the specific surface area of 1295.20 m^2/g, water content of 13.6%, ash content of 9.4%. Chemical activation with activating agent $ZnCl_2$ and physical activation on 850°C yields activated carbon with the specific surface area of 743 m^2/g, water content of 14.5%, ash content of 9.0%.

References

[1] Son H K, Sivakumar S, Rood, M J and Kim B J 2016 *J. Hzrd. Materials.* **301** 27-34
[2] Poulopoulos S G and Inglezakis V J 2006 *Adsorption, Ion Exchange and Catalysis: Design of Operations and Environmental Applications*: Elsevier Science
[3] Yuliusman 2015 Pembuatan Karbon Aktif dari Tempurung Kelapa Sawit dengan Bahan Pengaktif KOH dan Gas N_2/CO_2: *Proc Seminar Teknologi dan Rekayasa (SENTRA) (Universitas Muhamadiyah: Indonesia)* pp 78-84
[4] Hsu L Y and Teng H 2000 *Fuel Proc. Tech.* chapter 64 pp 155-166
[5] Teng H and Hsu L Y 1999 *Ind. Eng. Chem. Res.* **38** 2947
[6] Yacob A R, Siti Z H, Vicinisvarri I and Ratna S D 2009 Nano tungsten carbide prepared from palm kernel shell for catalytic decomposition of hydrazine: *Proc. Int. Conf. on Chemical, Biology and Environmental Engineering, CBEE (Singapore)*
[7] Khor KH, K O Lim and Zainal Z A 2009 *Amer. J. Appl. Sci.* **6** 1647
[8] Tsubouchi, Naoto, Nishio, Megumi and Mochizuki Y 2016 *Appl. Surf. Sci.* **371** 301
[9] Park K M, Nam H G, Lee K B and Mun S 2016 *J. Ind. Eng. Chem.* **34** 21-26
[10] Ramesh T, Rajalakshmi N and Dhathathreyan K S 2015 *J. Ener. Stor.* **4** 89-95

Removal of Mercury in Liquid Hydrocarbons using Zeolites Modified with Chitosan and Magnetic Iron Oxide Nanoparticles

E Kusrini[1], B H Susanto[1], D A Nasution[2], R Jonathan[1], W M Khairul[3]
[1]Department of Chemical Engineering, Faculty of Engineering, Universitas Indonesia, Kampus Baru UI Depok, 16424, Indonesia
[2]Center for Development of Agricultural Engineering, Agricultural Research and Development Agency, Ministry of Agriculture, Tangerang, Indonesia
[3]School of Fundamental Science, Universiti Malaysia Terengganu, 21030, Kuala Terengganu, Malaysia

E-mail: ekusrini@che.ui.ac.id, *Tel. 62-21-7863516 ext. 204, Fax. 62-21-7863515*

Abstract. Clinoptilolite zeolites were chemically modified with chitosan (Chit) and magnetic iron oxide nanoparticles (Fe_3O_4NPs) were synthesized for removal of mercury from liquid condensate hydrocarbon. The mercury content was in liquid hydrocarbon which was measured by Lumex mercury analyzer. The performance of sorbents based on zeolites modified chitosan and magnetic nanoparticles were examined on the real liquid condensate hydrocarbon. Removal of mercury using a prestine clinoptilolite zeolites, and zeolites modified chitosan (zeolites-Chit) were ~4.5, and ~35%, respectively. The effects of magnetic nanoparticles in zeolites-Chit sorbents were significant to reduce the mercury content in liquid condensate hydrocarbon which were from ~63 to ~66%. Increasing the mass ratio of Fe_3O_4 that influenced to the BET surface area of natural zeolites. Zeolites-Chit-Fe_3O_4NPs as an efficient sorbents are potential ideal to remove mercury in hydrocarbon for practical applications.

1. Introduction

Mercury presents in the condensate liquid and gaseous hydrocarbons in various forms, such as elemental mercury, mercury halide (inorganic mercury), complex mercury, mercury sulfide (suspended mercury) and organometallic compounds [1]. Mercury can be reacted with other metals to form amalgam. When this amalgam binds to the metal components in the equipments, it produces corrosion in some equipments. In the processing of gas, mercury can damage some equipments, especially in cryogenic heat exchanger, the doors, and the flow valve wellhead. In the petrochemical industry and oil refining, mercury can poison for catalyst and contaminate in wastewater. In addition, mercury is also harmful to human health [2].

Zeolites has been reported as adsorbent for removal of mercury [2]. On the hand, chitosan coated with magnetic nanoparticles for removal of mercury in aquoues and oily samples have been also reported by Nasirimoghaddam et al.[3]. As we know that the removal of mercury or reduction of metal content have been done by chemical precipitation, ion exchange, membrane, solvent extraction and adsorption [3]. Adsorption method was know as a highly effective and economical process for removal of mercury [2,3]. Magnetic iron oxide nanoparticles (Fe_3O_4NPs) is a material with strong magnetic properties, thus it can produce the active surface of magnetite and has a large surface area. Magnetic property can be applied in the field of environment, especially used as adsorbent to extract heavy metal ions [4]. Adsorbent having a magnetic properties can adsorp heavy metals from aquoues and oily samples, thus after adsorption process is carried out, the adsorbent can be separated from the medium by a simple magnetic methods [5-7].

In this research, zeolites was chemically modified with chitosan (Chit) and magnetic iron oxide nanoparticles (Fe_3O_4NPs) to produce adsorbents for removal of mercury in liquid hydrocarbon. The Fe_3O_4NPs was synthesized by co-precipitation method. The advantage of this method is it can be carried out at room temperature and easy to control the particle size, thus duration required is short [8]. The effects of chitosan and Fe_3O_4NPs in the surface of zeolites as sorbents were also studied.

2. Experimental

2.1. Materials
Natural clinoptilolite zeolites with moderate sizes (1 - 2 mm) were obtained from Lampung, Indonesia. Chitosan was purchased from Biotech Surindo [Cirebon, Indonesia]. $FeCl_2.4H_2O$, $FeCl_3.6H_2O$ and acetic acid were purchased from Merck. All chemicals used were analytical grade. The real liquid condensate hydrocarbon was provided from Sumatera, Indonesia.

2.2. Preparation of magnetic iron oxide nanoparticles (Fe_3O_4NPs)
Magnetic iron oxide nanoparticle was prepared by co-precipitation method. The mass of $FeCl_2.4H_2O$ and $FeCl_3.6H_2O$ were varied with ratio 2:1 (w/w), namely 1.0 g $FeCl_2.4H_2O$ and 2.6 g $FeCl_3.6H_2O$ (1), 1.5 g of $FeCl_2.4H_2O$ and 3.9 g $FeCl_2.4H_2O$ (2), and 2.0 g $FeCl_3.6H_2O$ and 5.2 g $FeCl_3.6H_2O$ (3). Each of the mixture of $FeCl_2.4H_2O$ and $FeCl_3.6H_2O$ were disolved in 10.3 mL HCl 1 N and added 15 mL distillated water, then the mixture was stired. The solution was poured slowly into 250 mL1.5 M NaOH solution, and keep stiring for 1 h. The solution was separated by magnetic and the precipitate was washed with distilled water for several times. The dark brown precipitate of magnetic iron oxide nanoparticles (Fe_3O_4NPs) was obtained and heated at 60°C for overnight and followed dried at 100°C for 2 h.

2.3. Preparation of zeolites-Chit
The synthesis of zeolites-Chit as sorbent was modified from the combination methods according to procedures reported by Kusrini et al. [9]. The activated clinoptilolite zeolites was coated with chitosan solution in acetic acid 1% (v/v), then the mixture was stirred for 4h, followed was filtrated by vacuum filter. The zeolites-chit sample was dried at 60-70°C for 4 h in commercial oven.

2.4. Preparation of zeolites-Chit-Fe_3O_4NPs
Each of Fe_3O_4 composition was mixed with activated clinoptilolite zeolites-chitosan solution 50 mL, then it was stired for 4 h and then the mixture was dried at 60°C for 24 h. Three types of new adsorbents of zeolites-Chit-Fe_3O_4 were characterized by FTIR, SEM-EDX, and BET.

2.5. Adsorption test to remove of mercury in liquid condensate hydrocarbon
All adsorption experiment to removal of mercury in liquid hydrocarbon were conducted at room temperature. A batch experiment was applied to investigate the effect of adsorbents. The adsorbent dose was selected only for 0.5 g, volume of the liquid hydrocarbon of 30 mL, then the mixture was stired for 2 h and it kept for 24 h to enhance the adsorption process. After the eqilibrium was achieved, the suspension was filtered using a vacuum filter. The mercury levels of liquid hydrocarbon after treated with adsorbents was analyzed using a mercury analyzer.
For comparison purpose, the adsorption of mercury by unmodified zeolites (a prestine of zeolites), zeolites-Chit and zeolites-Chit- Fe_3O_4NPs have been conducted. The effect of types of adsorbents on the removal of mercury content before and after adsorption were compared. Amount of mercury per

gram of adsorbent was calculated based on differences in initial and final concentrations of mercury after adsorption, with equation (1) and (2) as follows:

$$q_e = \frac{M\,(C_o - C_e)}{W} \quad (1)$$

$$\eta = (Co - \frac{C_e}{C_o}\ x\ 100\% \quad (2)$$

Here, η is adsorption efficiency of mercury, q_e is the amount of mercury adsorbed in adsorbent (ng/g), C_0 and C_e are initial and equilibrium concentration of mercury (ng/g), respectively, M is the number of liquid condensate that used (g), and W is the mass of adsorbent in the measurements (g).

2.6. Characterization

Characterization of unmodified zeolites, zeolites-Chit and new adsorbents of zeolites-Chit- Fe_3O_4NPs were further identified by Fourier-transform infrared (FTIR; Bruker), scanning electron microscope and energy dispersive X-ray analyzer (SEM-EDX; Genesis). FTIR to identify the chemical bonding and the functional groups of the samples. The BET surface area and pore size distribution of all samples were measured through N_2 sorption analysis using ASAP 2020 (V4.02) Micromeritics instruments (USA). The concentration mercury content in liquid hydrocarbon was analyzed by Mercury Analyzer (Lumex, RA915).

3. Results and Discussion

3.1. FTIR studies

Characterization test was carried out to the adsorbent before and after modification in order to able to compare successful or not adsorbent made. The FTIR spectrum of the unmodified zeolites showed the absorption peaks at 470 and 792 cm^{-1} are assigned for the bending vibration absorption of T–O and and the absorption vibration of O-Si–O (see Figure 1(D)). The absorption peak at 1072 cm^{-1} showed absorption vibration of Si-O and Al–O. While the peak at 3620 cm^{-1} showed absorption vibration of hydroxyl (O –H), Si - OH and Si-OH-Al.

The FTIR spectra of zeolites after modified with chitosan and magnetic iron oxide nanoparticles was shown in Figure 1. Several new absorption peaks was observed after modification with chitosan and Fe_3O_4NPs. Two new absorption bands at 2891 cm^{-1} and 1612 cm^{-1} are assigned for the C-H and amine (NH_2) groups from chitosan were observed in the FTIR spectrum of zeolites-Chit (see Fig. 1(E)). After zeolites-Chit was chemically modified with Fe_3O_4, the new absorption peaks at 572 cm^{-1}, 601 cm^{-1} and 637 cm^{-1}, which indicates the Fe – O streching bands were observed (see Fig. 1(A-C)). It is confirmed that zeolites was successfull chemically modifed with chitosan and Fe_3O_4. The absorption peaks at 570 – 640 are usually assigned to the Fe-O bands [6].

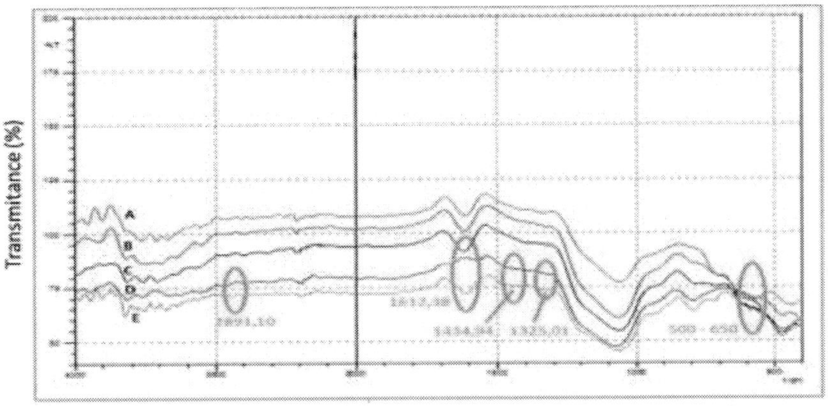

Figure 1. FTIR spectra of prestine zeolites (D), zeolites-Chit (E), zeolites-Chit-Fe_3O_4NPs (5.2:2 (w/w) (C), zeolites-Chit-Fe_3O_4NPs (2.6.:1 (w/w) (B), zeolites-Chit-Fe_3O_4NPs (3.9.:1.5 (w/w) (A)

3.2. Morphology and composition characterization

The SEM morphology of prestine zeolites, zeolites-Chit and zeolites-Chit-Fe_3O_4NPs was shown in Figure 2 (A-C). It was clear observed that changing in the surface of zeolites indicating chitosan and Fe_3O_4NPs are presence in the surface of natural zeolites. The white spot was observed in Fig.2(C) which indicate the presence of iron metals ion surface of zeolites.

A

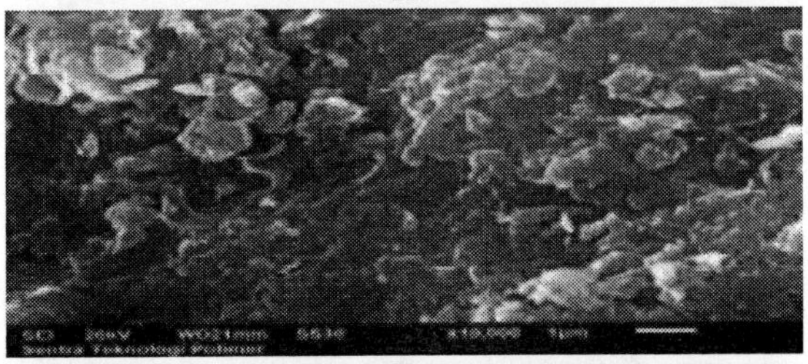

B

C

Figure 2. Images morphology of zeolites (A), zeolites-chitosan (B), and zeolites-Chit-Fe_3O_4 (C) with magnification 10,000x.

The EDX compositions of all adsorbents were summarized in Table 1. Increasing the mass of Fe in the zeolites-Chit-Fe_3O_4NPs adsorbent, which indicates that modification of the zeolites-Chit with Fe_3O_4NPs was successful. In addition, the Fe content in the unmodified zeolites showed the impurities in zeolites.

Table 1. Element composition of zeolites, zeolites-chitosan and zeolites-chitosan-Fe_3O_4NPs

Element	Composition (% weight)		
	Zeolites	Zeolites-*Chit*	Zeolites-*Chit*- Fe_3O_4NPs
C	4.66	10.60	10.15
O	51.29	49.29	47.61
Na	-	1.99	1.51
Al	6.27	6.11	6.01
Si	34.61	28.54	29.72
K	0.89	0.73	0.81
Ca	0.21	0.41	0.38
Fe	2.07	2.33	3.81

3.3. Surface area characterization

The surface area, nanoparticle size and adsorption pore size of adsorbents was sumarized in Table 2. After modification with chitosan and Fe_3O_4NPS, the surface area of zeolites-Chit-Fe_3O_4NPs adsorbents was increased and reached to 119.5 m^2/g. Increasing the mass ratio of Fe_3O_4 that influenced to the BET surface area of natural zeolites. The nanoparticle size of sorbent was significantly reduced from 93.2 to 50.2 nm after chemically modified with chitosan and Fe_3O_4NPs. It is confirmed that the presence all different mass ratio of Fe_3O_4NPs that effected the surface and size of sorbents. Adsorption average pore width (4V/A by BET) (nm) for all mass variation of zeolites-Chit-Fe_3O_4NPs are increased. The surface area of activated zeolites increased after the modification, thus we expected that the ability of zeolites-Chit-Fe_3O_4NPs sorbents to remove of mercury in liquid hydrocarbon also increased. Surface area for clinoptilolite zeolites is 64.4 m^2/g is higher than found in the clinoptilolite zeolites as reported before in our study (47.1 m^2/g) [9].

Table 2. BET characterization for zeolites, zeolites-Chit, zeolites-Chit-Fe_3O_4NPs

Adsorbent	BET Surface Area (m^2/g)	Nanoparticle size (nm)	Adsorption average pore width (4V/A by BET) (nm)
Zeolites	64.4	93.2	6.7
Zeolites-*Chit*	80.5	74.5	6.9
zeolites-Chit-Fe_3O_4 (2.6.:1 (w/w)	97.7	61.4	10.6
zeolites-Chit-Fe_3O_4 (3.9.:1.5 (w/w)	111.3	53.9	11.3
zeolites-Chit-Fe_3O_4 (5.2:2 (w/w)	119.5	50.2	10.0

3.4. Removal of mercury in the real liquid condensate hydrocarbon

The mercury adsorption performance test was conducted in batch experiment. The effect of chemical modification for activated clinoptilolite zeolites sorbent with chitosan and Fe_3O_4NPs to adsorb mercury in liquid hydrocarbon was studied. From the Table 3, it was clear that the mercury levels in ng/g (ppb) after treatment with zeolites, zeolites-Chit and zeolites-Chit-Fe_3O_4NPs adsorbents shown on mercury concentrations in pbb. The concentration of mercury in the real liquid condensate hydrocarbon before adsorption is 101 ppb. After treatment with zeolites, the mercury content is reduced to be 96.5 ppb. The presence of amine and hydroxyl groups from chitosan in the zeolites-Chit sorbents was significantly reduced and reached until 65.5 ppb. This chemical bonding is due to the incorporation between mercury metal ions with the functional groups of amine (NH_2) and hydroxyl (O-H) coming from chitosan. The removal of mercury using an activated clinoptilolite zeolites, and zeolites-Chit were ~4.5, and ~35%, respectively.

Table 3. Removal of Mercury in Liquid hydrocarbon

Type of adsorbents	Concentration (ppb)	Removal of mercury (%)
Liquid hydrocarbon (before treatment)	101	
Zeolites	96.5	4.5
Zeolites-Chit	65.5	35.1
zeolites-chit-Fe_3O_4NPs (2.6.:1 (w/w)	37.5	62.9
zeolites-chit-Fe_3O_4NPs (3.9.:1.5 (w/w)	36.0	64.4
zeolites-chit-Fe_3O_4NPs (5.2:2 (w/w)	34.0	66.3

The adsorption of mercury by zeolites-Chit-Fe_3O_4NPs was investigated at three different mass ratio of Fe_3O_4NPs and the data adsorption was analyzed by mercury analyzer. The presence of Fe_3O_4NPs in the zeolites-Chit has a significant effec to remove the mercury content in the real liquid hydrocarbon. The effects of magnetic nanoparticles (Fe_3O_4NPs) in zeolites-Chit sorbents were significant to reduce the mercury content in liquid condensate hydrocarbon which were from ~63 to ~66%.

The findings from characterization of zeolites-Chit-Fe_3O_4NPs adsorbent showed that the adsorbent was successfully modified by the addition of Fe_3O_4NPs compared with the adsorbent prior to modification. Based on the results, it can be seen that the modifications were made to improve the adsorption properties on the adsorbent, wherein the adsorption with only ~ 4.5% by prestine zeolites and zeolites-Chit with only ~35.1% of mercury were absorbed. While the adsorption zeolites-Chit-Fe_3O_4NPs adsorbent gave ~66.3% of mercury absorbed.

4. Conclusion

In this paper, the modification of natural zeolites with chitosan and Fe_3O_4NPs were prepared by impregnation method. The utilization of zeolites-Chit-Fe_3O_4NPs as adsorbent for removal of mercury in liquid hydrocarbon were investigated. The experimental results had shown that the modified zeolites with chitosan and Fe_3O_4NPs (zeolites-chit-Fe_3O_4NPs) were effective in reducing and remove of mercury in liquid hydrocarbon by removal efficiency are about 63 – 66%.

Acknowledgments

We acknowledge Universitas Indonesia as financial support through the grant *Hibah Kompetensi Untuk Publikasi di Jurnal Internasional Terindeks*, No. 2529/UN2.R12/HKP.05.00/2016 and KKP3N, No. : 54.30/HM.240/I.1/3/2016.K, a project funded by the Ministry of Agriculture, as provider for zeolites.

References

[1] Wilhelm S 1999 Conceptual Design of Mercury Removal System for Hydrocarbon Liquids, Hydrocarbon Processing, 233-263
[2] Chojnacki A, Chojnacka K, Hoffmann J and Gorecki H 2004 *Minerals Eng.* **17** 933–937
[3] Nasirimoghaddam S, Zeinali S, and Sabbaghi S 2015 *J. Ind. Eng. Chem.* **27** 79–87
[4] El Ghandoor H, Zidan H M, Khalil M M H and Ismail M I M 2012 *Int. J. Electrochem. Sci.* **7** 5734-5745
[5] Shen X, Wang Q, Chen W L and Pang Y 2014 *App. Surf.Sci.* **317** 1028–1034
[6] Parham H, Zargar B, and Shiralipour R 2012 *J. Hazard.Mat.* **205– 206** 94– 100
[7] Ghasemi E, Heydari A and Sillanpää M 2017 *Microchem. J.* **131** 51–56
[8] López R M G, Hurtado P G, de León R D, Fernández S, Saade H and Bueno D 2013 *Int. J. Mol.Sci.* **14(10)** 19636–19650
[9] Kusrini E, Lukita M, Susanto B H, Nasution D A, Rahman A, and Gunawan C 2016, Draft for preparation.

IOP Conf. Series: Earth and Environmental Science **75** (2017) 012011 doi:10.1088/1755-1315/75/1/012011

Seepage Zone Identification at Sutami Dam by Means of Geoelectrical Resistivity Data

Sunaryo [1], **A. Susilo**[1]
[1]Program Study of Geophysics, Department of Physics, Faculty of Sciences, University of Brawijaya, Malang, Indonesia

E-mail: sunaryo@ub.ac.id; sunaryo.geofis.ub@gmail.com

Abstract. A study with entitled Seepage Zone Identification At Sutami Dam By Means Of Geoelectrical Resistivity Data has been conducted. The aim of this study is to get the distribution of seepage zones in the Sutami dam body. Data acquisition has been done by using Oyo-EL McOhm Model 2119C equipment on the tracks along the side of the dam body, both downstream (exhaust) side for first and upstream (reservoir) side for second with pseudodepth section dipole-dipole configuration. Each track has a length of 400m and 380m respectively. The distance between the point electrode consists of 1 track with space (a) is 10m (n= 8) and 3 tracks with a space (a) is 20m (n= 8). Based on the results of data processing obtained are low resistivity values ($\rho \leq 9.57\Omega m$), the moderate resistivity value ($9.57\ \Omega m < \rho \leq 320\Omega m$), and high resistivity ($\rho > 320\Omega m$). The distribution of fracture zones associated with seepage zones for Sutami dam distributed over the surface to depth of 8-18 m, both upstream (reservoir) or downstream (exhaust) indicated by resistivity values 0.922 Ωm to 9.57 Ωm. The results of the 3D data processing indicates that the fracture zone is also distributed between the upstream (reservoir) and downstream (exhaust) of the Sutami dam body.

1. Introduction

Sutami dam is one of part of the Karangkates dam where is located at the village of Karangkates, Sumberpucung district, Malang regency. Karangkates dams consists of two parts, i.e. Lahor dam (inaugurated in 1977) and Sutami dam (inaugurated in 1981). Besides as hydro power plant, Karangkates dams also as a tourist spot, control of flood and irrigation. Karangkates dams able to contribute power of about 400 KWH per year from total need about 29.373 MW installed in all Indonesia area in the last five years [1].

Hypothetically, Karangkates hydro power plant located adjacent to the area of contact between the andesite lava and limestone as well as shear fault which is located at around Pohgajih area [2]. These geological conditions caused in local site becomes unstable compared to the surrounding. Some findings out crop of this condition has been found in the type of cracks along the rail road and cracks and subsidence of the road which is located around the dams. For the case occurred in the village of Karangkates because of a shift or movement is great enough land around the year 1999. It was about 78 families, forced to flee to house of Jasa Tirta Transito for about 2 weeks. Cause of the movement of the surface, almost all houses of Ledok village cracks [3], [4]. Indications of the cracks also occurred at around of the railway line [5].

[1]Author for correspondence.

The general objectives to be achieved in this research is the safety condition of the dams in order to support energy sustainability. Three important things as reasons for the need of this research are: 1). The age of the Sutami dam was 35 years old, 2). The location of the dams are closed together to the Pohgajih local shear fault, Selorejo local fault, and Selorejo limestone-andesite rocks contact plane and 3). Karangkates dams is one of the important hydro power plant PLTA in Indonesia. Geoelectrical resistivity used for detecting the impact of geological activity on the surface (near surface) at the body of Sutami dam as follow up of the previous research, i.e: Vulnerability of Karangkates dams area by means of density contrast parameter to anticipate energy sustainability [6] and Vulnerability of Karangkates dams area by means of zero crossing analysis of data of magnetic [7].

2. Experimental

This research carried out by using Oyo McOhm-EL Model 2119C instrumentation. Geoelectrical resistivity data acquisition performed on the path along the right and left side of the dam body by using pseudodepth section dipole-dipole configuration [8]. (figure 1)

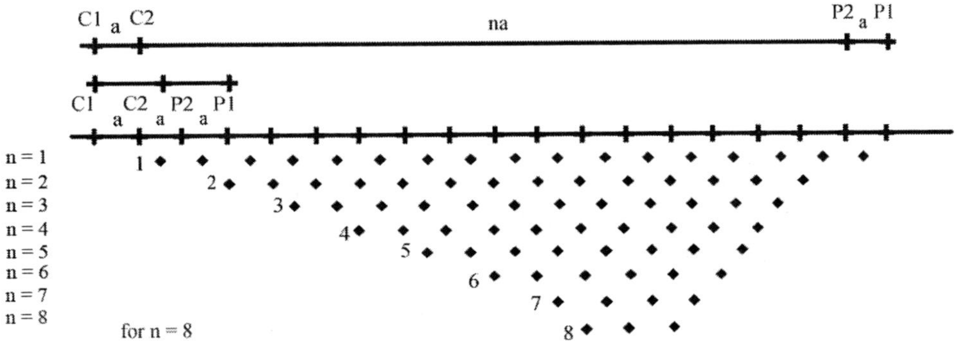

Figure 1. The configuration arrangement of pseudodepth section.

Apparent resistivity values of geoelectrical resistivity method with pseudodepth section dipole-dipole configuration mentioned above are as equation (1).

$$\rho=k(\Delta V/I), \text{ with: } k=\pi n(n+1)(n+2)a \quad (1)$$

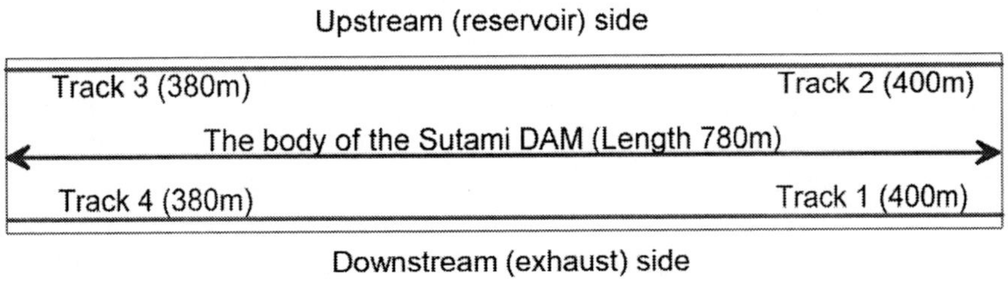

Figure 2. Tracks of geoelectrical resistivity data acquisition at Sutami dam.

Tracks of data acquisition consists of 2 tracks on the left side of the dam, and 2 tracks on the right side of the dam. Each track has a length of 400m and 380m. The spacing between the electrode consists of 1 track with a space (a) is 10m (n= 8) and 3 tracks with a space (a) is 20m (n= 8) (figure 2).

Data processing and interpretation for geoelectrical resistivity method with dipole-dipole pseudodephsection configuration using Jacobian matrix and Quasi-Newton method [9].

3. Results and Discussion

Geoelectrical resistivity data acquisition at Sutami dam are 4 tracks of data acquisition with parallel arrangement between tracks. Two tracks are in the body of the upstream dam (reservoir) side and 2 tracks are in the body of the downstream dam (exhaust) side.

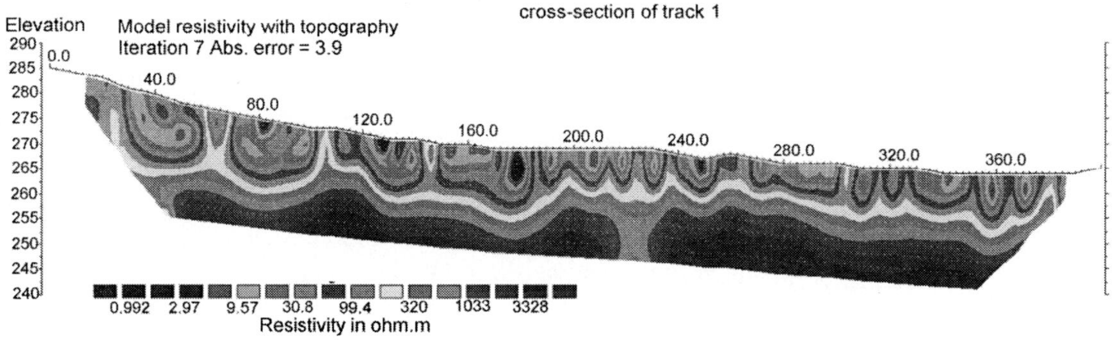

(a)

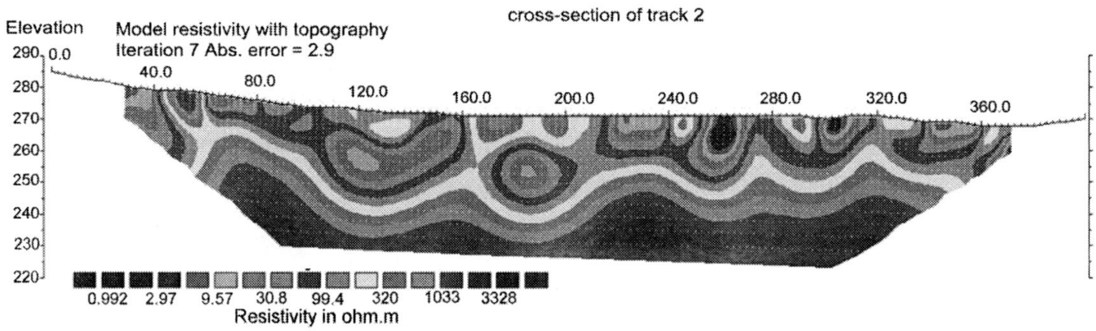

(b)

Figure 3. (a). 2D cross-section of track 1 (downstream side / exhaust of Sutami dam), (b). 2D cross-section of track 2 (upstream / reservoir of Sutami dam).

Tracks 1 is a track where is at the body of the dam on the downstreem dam with track length is 400m, spacing between electrode is a=10m and n=8. Tracks 2 is a track where is at the body of the dam on the upstream dam with track length is 400 m, spacing between electrode is a=20m and n=8. Both tracks are parallel each other and data acquisition performed from the northeast to the southwest.

Based on the visual observation there are some cracks are visible on the sidewalk side of the downstream as well as cracks at the road body at the top of the Sutami dam. Geoelectrical resistivity data processing results show a pattern layer on each track which is then interpreted by the variation of the resistivity of the subsurface rock. Figure 3 shows a cross section 2D for track 1 (the upstream) and track 2 (downstream) from the results of data processing by using software RES2DINV.

Figure 3 (a) shows a 2D cross section of track 1 (downstream). 2D cross section shows the subsurface conditions of the track 1, where blue indicates the existence of sandstone with water contents, green to light brown indicate sandstone with no water contents and orange to dark brown color shows the sand. Figure 3 (b) shows a 2D cross section of the track 2 (the upstream), which under the condition of the track surface has the same resistivity variations with track 1 (downstream).

Based on the results of the visual observation shows that there are cracks at some points in the downstream (exhaust) among others at a point 20 m (figure 4 (a)), 45 m (figure 4 (b)), 55 m (figure 4 (c)), and 65 m (figure 4 (d)). Cracks are visible in the pavement that is in the body of the downstream (exhaust) of the dam side. Cracks were mostly right on the part of low resistivity are shown by the blue color. This shows that the zones with low resistivity (blue) indicates the fracture zones that associated with seepage zones in track 1.

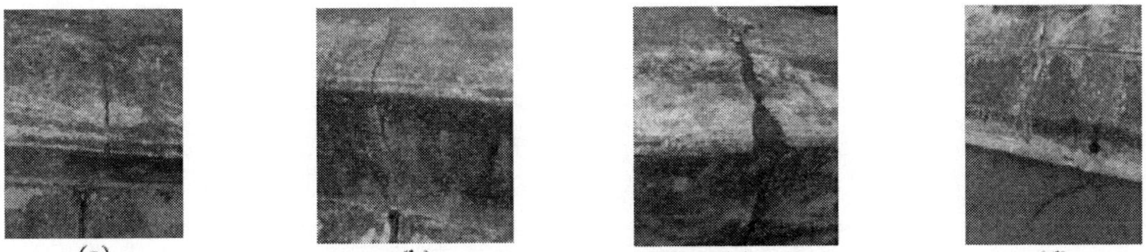

(a) (b) (c) (d)

Figure 4. Visual observations on the dam body at the downstream (exhaust) of Sutami dam for track 1.

To combine the two interpretations of the track then done using sotfware Geosoft Oasis Montaj order to obtain a 3D cross-section. Figure 5 shows a cross section based 3D resistivity geoelectric data processing for track 1 (downstream/exhaust side) and track 2 (the upstream/reservoir side).

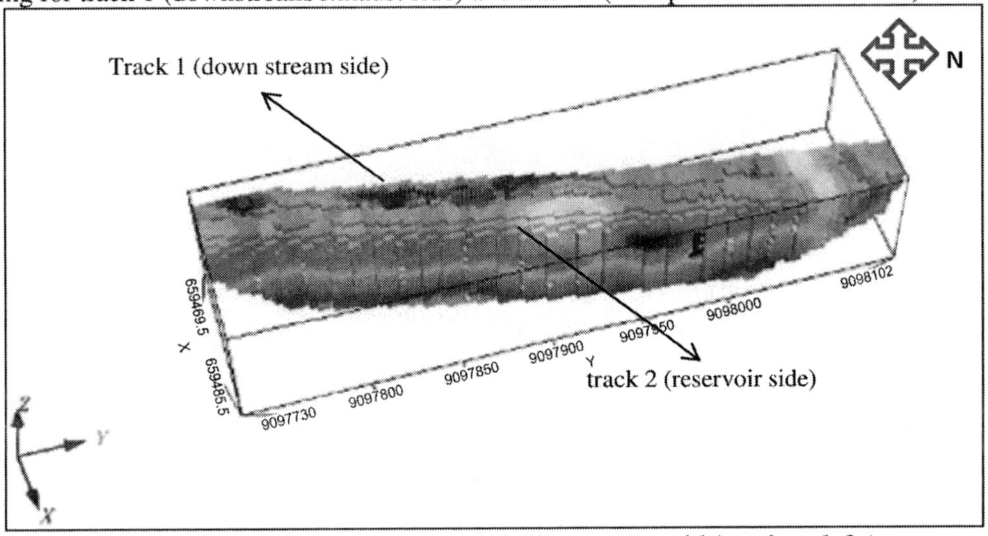

Figure 5. 3D cross-section of the combined track 1 (downstream side) and track 2 (upstream side).

Based on figure 5 shows that the zones with low resistivity are spread more visible at the downstream side that represented by the blue color. This low resistivity indicates that there are distribution of fracture zones that associated with seepage zones at the downstream side. It is seen also on the results visually, which is commonly found cracks points at the downstream side. Low resistivity spread on the top to depth of about 8 m from the top of the Sutami dam body.

Track 3 is a track at Sutami dam on the upstream (reservoir) side which is a continued from track 2. Track 3 (upstream / reservoir side) has a track length of 380 m with length spacing is a=20m and n=8. There is also track 4 which is located at the downstream (exhaust) side which is a continued from the track 1. The track 4 (downstream / exhaust) side has length track is equal to track 3 as well as the use of space and n the same too. Both tracks are also parallel configuration each other and data acquisition performed from the northeast to the southwest.

On the sidewalk was on the road body at downstream (exhaust) side, visually also seen many cracks as on the downstream (exhaust) for track 1 has been described previously. The results of the geoelectrical resistivity data processing for 2D modeling using RES2DINV software for track 3 (the upstream / reservoir side) and the track 4 (downstream / exhaust side) are shown in figure 6.

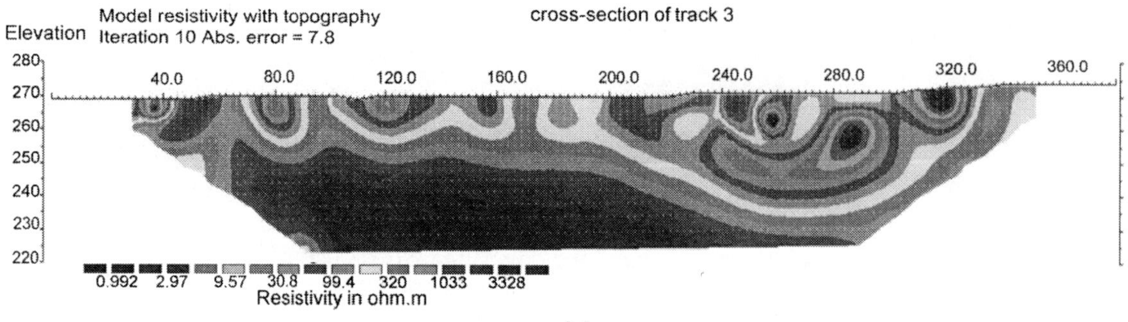

(a)

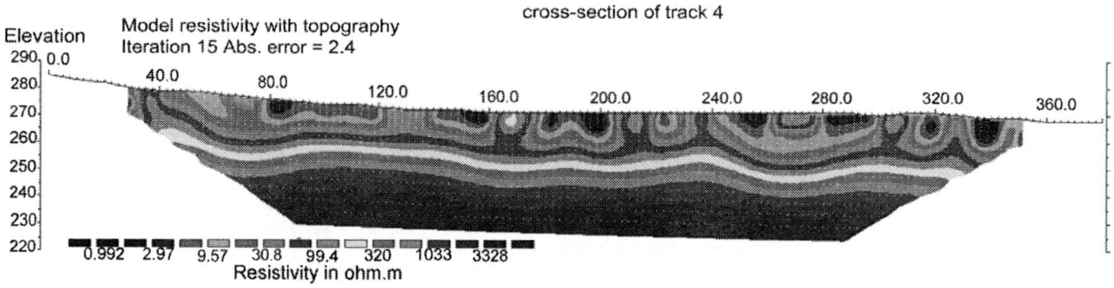

(b)

Figure 6. (a). 2D cross-section of track 3 at upstream (reservoir) side, (b). 2D cross-section of track 4 at downstream (exhaust) side.

Figure 6 (a) shows the subsurface conditions of the track 3 (upstream side), which subsurface conditions have variations in the resistivity values equal to track 1 (upstream side) and track 2 (downstream side). The difference is in the distribution of resistivity values, which for the low resistivity only a small parts visible in track 3 (upstream side). Figure 6 (b) shows the subsurface conditions of the track 4 (downstream side). At track 4 (upstream/reservoir side) showing low resistivity zones spread across the top to a depth of about 12 m. The zone of low resistivity indicates the weak zone which in this case is a fracture zones associated with seepage zones. Based on visual observation was not visible cracks in the pavement section upstream/reservoirs side, but seen some cracks in the downstream side among others located at a point 5 m (figure 7 (a)), 115 m (figure 7 (b)), 140 m (figure 7 (c)), 222.5 m (figure 7 (d)), and 315 m (figure 7 (e)). Cracks that if correlated to the track of 2D cross section at track 4 (downstream side) shown that almost the entire position of the cracks are in the blue zone or low resistivity. This shows that the fracture zone can be interpreted as having low resistivity zones associated with the seepage zones of Sutami dam.

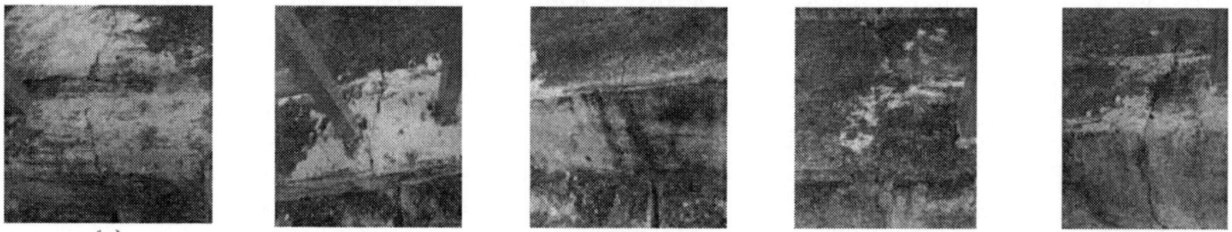

(a) (b) (c) (d) (e)

Figure 7. Visual observations results on the downstream of Sutami dam at track 4.

To combine between track 3 and the track 4 use Geosoft Oasis Montaj software that gives 3D picture. Figure 8 shows a cross section of the 3D results of the resistivity geoelectric data processing.

Zones with low resistivity is spread from the top to a depth of approximately 12 m. In the figure 8 shows that the zones with low resistivity distributed at downstream side, i.e at track 4. This indicates conformity with the results visually indicating cracks that occurred some point at downstream side. Zones with low resistivity are distributed from the top to the depth of approximately 12 m.

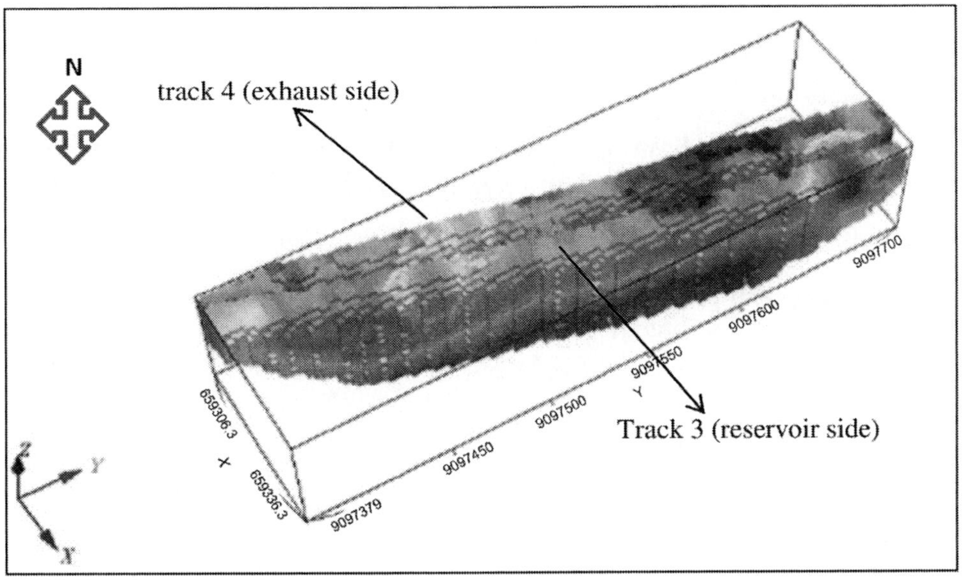

Figure 8. 3D cross-section of the combined track 3 (upstream/reservoir side) and track 4 (downstream/exhaust side).

4. Conclusions

Based on the results of the research using geoelectrical resistivity method at Sutami dam can conclude that from the results of data processing obtained the low resistivity values ($\rho \leq 9.57\Omega m$), moderate resistivity values ($9.57\ \Omega m < \rho \leq 320\Omega m$), and high resistivity ($\rho > 320\Omega m$). The distribution of fracture zones associated with seepage zones for Sutami dam distributed over the surface to the depth of 8-18 m, both upstream or downstream indicated by resistivity values 0.922 Ωm to 9.57 Ωm. The results of the 3D data processing of geoelectrical resistivity data indicates that the fracture zone is also distributed between the upstream and downstream from the body of the Sutami dam.

References

[1] http://www.engineeringtown.com 2013 access date 10 Maret 2013
[2] Sjarifudin M Z and Hamidi S 1992 *Peta Geologi lembar Blitar, Jawa skala 1:100*.000 (Bandung: Pusat Penelitian dan Pengembangan Geologi, Direktorat Jenderal geologi dan Sumberdaya Mineral, Departemen Pertambangan dan Energi)
[3] Malang post 2009a *Kampung Ledok Dihantui Tanah Bergerak* Malangpost online 1 February 2009
[4] Malang post 2009b *Waspadai Bendungan Karangkates* Malangpost online 30 March 2009
[5] Susilo A, Sunaryo and Wasis 2013 *Journal of Traffics and Logistics Engineering (JTLE)*
[6] Sunaryo and Susilo A 2015 Vulnerability of Karangkates Dams Area By Means Of Density Contrast Parameter to Anticipate Energy Sustainability: *Proc. The 5th Annual Basic Science International Conference*
[7] Sunaryo and Susilo A 2015 *American International Publishing (AIP)* **1658**
[8] Loke M H and Barker R D 1996 *Geophysics Prospecting* **44** 131-152
[9] Loke M H 2000 *Electrical Imaging Surveys for Environmental and Engineering Studies: A Practical Guide to 2-D and 3-D Survey* Penang Malaysia

Effect of Light Intensity for Optimum Biomass and Lipid Production from *Scenedesmus dimorphus* (Turpin) Kützing

F N Kurniawati[1], E Mahajoeno[1], Sunarto[1], and S L A Sari[1]
[1] Program of Study Biology, Faculty of Mathematics and Natural Sciences, Sebelas Maret University, Surakarta

email: edmasich@yahoo.com

Abstract. One source of alternative energy substitute for petroleum raw materials is renewable vegetable oils known as biodiesel. Biodiesel can be produced from microalgae, since it was more efficient and environmentally friendly. *Scenedesmus dimorphus* (Turpin) Kützing was developed as a source of biodiesel since it had potential of high lipid production. The aims of this research were to know the rate of growth of *Scenedesmus dimorphus* in different lighting and the optimimum light intensity for biomass and lipid production. This research used a completely randomized design consisting of 3 treatments with 3 replications. Treatments in this research were the light intensity, i.e. 7,500, 10,000, and 12,500 lux. *Scenedesmus dimorphus* was grew in Bold's Basal Medium (BBM). Parameters observed in this research were the cell number, biomass and lipid production of *S. dimorphus*. Data were analyzed by ANOVA followed by DMRT 5%. The results showed that the optimum growth rate of *S. dimorphus* was in the intensity of 12,500 lux that was 100.80 x 10^6 cells.ml^{-1}. The optimum production of biomass and lipids was in treatment 12,500 lux i.e; 1.1407 g.L^{-1} and 0.2520 g.L^{-1} (22.28% dry weight).

1. Introduction

Energy is needed in the lives of both humans and other organisms. Energy derived from the natural resources of renewable and non-renewable. Innovation of alternative energy sources start to be developed considering the stock of non-renewable energy sources such as oil decreases. Alternative sources of energy are used as a replacement of petroleum feedstock is biodiesel. The use of biodiesel as a diesel substitute standard will not only help the environment, but also will help increasing energy independence and energy security of the country. Biodiesel can be produced from vegetable oils. Microalgaee are one source of oil that can be used for making biodiesel, because microalgaee contain a high enough lipid.

Lipids in the cell of microalgaee serve as a backup energy source in cells deficient in carbohydrates as an energy source. Another function of lipid microalgaee can be utilized as a source of alternative energy as a raw material for biodiesel, biodiesel privilege that comes from microalgae i.e. can be updated (renewable), non-toxic, and can be broken down naturally (biodegradable) [1]. The processes of the biosynthesis of lipids on microalgae require great energy, begin with the process of the formation of carbohydrates then continue with the accumulation of lipids [2]. A microalgae that can be developed as a source of biodiesel is *Scenedesmus dimorphus* (Turpin) Kützing. *S. dimorphus* is a comsopolit microalgae, mostly live in the aquatic neighborhood, in fresh water or brackish water. *Scenedesmus dimorphus* act as producers in the ecosystems as oxygen providers [3].

Energy generated in the process of photosynthesis can be used as growth in microalgae, spare food or to defend themselves when the pressure on the environment [4]. The autotroph organisms Microalgae are capable to form organic compounds from inorganic compounds through the process of photosynthesis. The existence of the light determines the curve shape of growth for microalgae that performs photosynthesis. [5] do the culturing microalgae *S. dimorphus* in different variations of the Bold's Basal Medium (BBM) shines with 10,000 lux shows that the highest biomass production are present in 125% of BBM media 0.42 g/l, while the production of lipids found in the highest media 200% of BBM 0.026 g/l (18.49% dry weight).

But in fact, until now, especially in light of microalgae utilization as energy source is still not optimal. From the description above, it needs to be carried out a research on the influence of the light shines towards the acceleration of growth in microalgae biomass and lipids to produce the optimum total.

2. Experimental

2.1 Microalgae strain and cultur medium

A microalga used in this research is the culture of Scenedesmus dimorphus (Turpin) Kützing obtained from Laboratory Research Center for Limnology-LIPI Cibinong. Bold's Basal Medium (BBM) was used to grow the pure culture of *S. dimorphus*. The stock medium consisted of (g L^{-1}): $NaNO_3$ (25.0 g), $CaCl_2.2H_2O$ (2.5 g), $MgSO_4.7H_2O$ (7.5 g), K_2HPO_4 (7.5 g), KH_2PO_4 (17.5 g), NaCl (2.5 g), EDTA (50.0 g), KOH (31.0g), $FeSO_4.7H_2O$ (4.98g), H_2SO_4 (1.0 ml), H_3BO_3 (11.42 g), $ZnSO_4.7H_2O$ (8.82 g), $MnCl_2.4H_2O$ (1.44 g), MoO_3 (0.71 g), $CuSO_4.5H_2O$ (1.57 g).

2.2 Cultivation conditions

The microalgae was cultivated in laboratorium at 28-30°C. Various operating conditions were assayed including light intensities of 7,500 lux, 10,000 lux, and 12,500 lux. The light intensity were measured with a light meter. Temperature and pH of the culture medium were monitored daily.

2.3 Analytic determinations

2.3.1 Cell number, biomass, and lipid concentration

The cell concentration of culture was determined regulary by measuring optical density at wavelength 680 nm (denoted as OD_{680}) using a UV/VIS spectrophotometer and converted into a number of cells with the regression curve. Extraction of lipids was done on the day when the first appropriate exponential growth and density measurement results cells. As many as 50 ml culture tube centrifuge microalgae was centrifuge with the speed of 3,000 rpm for 5 minutes. The pellets then dried with an oven for 24 hours at a temperature of 80°C and weighed as a dry weight. The dry Pellet is suspended with the solvent chemical that consists of 2 ml of distilled water, ion-free 5 ml methanol and 2.5 ml of khloroform and then be homogened with the help of shaker horizontally for 24 hours. Into the results stirring 2.5 ml water added back destilat free ion and 2.5 ml of khloroform then do a speed 4,000 rpm sentrifuse for 25 minutes. A mixture of lipids and chloroform placed separately from the other materials then be heated so that chloroform can be evaporated, the rest mixture are dried and weighed as lipid weight of lipids. Percentage of lipids can be calculated with the formula [6] :

$$\%\ \text{Lipid} = \frac{\text{Lipid Weight}}{\text{Dry Weight Microalgae}} \times 100\ \% \qquad (1)$$

While the productivity of lipids can be calculated with the formula [2] :

$$\text{Productivity (g/l/days)} = \frac{\text{Lipid Weight / l}}{\text{The Number Of Days}} \qquad (2)$$

3. Result and Discussion

3.1 The growth pattern of S. dimorphus

Growth curve of *S. dimorphus* (Figure 1) shows a high light intensity effect on the growth. The highest cell numbers appears on the light intensity 12,500 lux with 100.8 x 10^6 cell.ml^{-1}, and the amount of the lowest cell appears at 10,000 lux light intensity with 65.3 x 10^6 cell.ml^{-1}. While the culture of 7,500 lux is in the middle with a cell number 70.8 x 10^6 cell.ml^{-1}. The higher intensity light is used, the higher the microalgae cell growth. The intensity of the light is indispensable in the process of photosynthesis because it relates to the amount of energy received by the microalgae to perform photosynthesis.

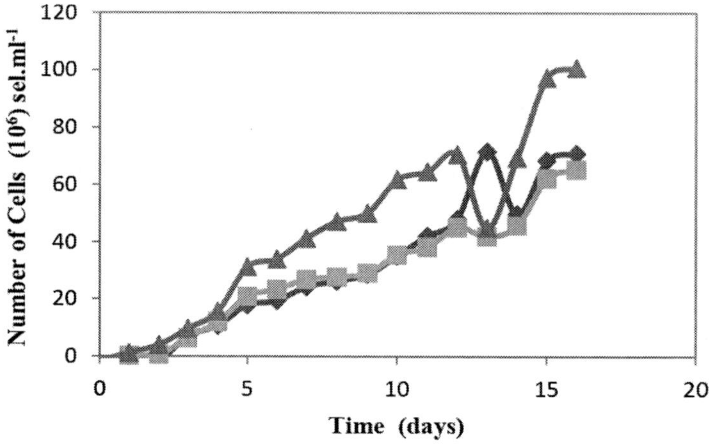

Figure 1. Growth curve of *S. dimorphus* in different light intensity 7,500 lux (—◆—), 10,000 lux (—■—), and 12,500 lux (—▲—)

Light is the main energy source for *S. dimorphus*. If the supply of light is limited, the supply of nutrients will decrease. Thus, *S. dimorphus* growth will be hampered. According to [7], the light is needed in the process of photosynthesis to produce energy (hexoses) and oxygen from carbon dioxide and water. Energy from light is used in the electron transfer process to CO_2 through the reduction of nicotinamide adenine dinucleotide phosphate (NADP) becomes NADPH. Electrons are extracted from the water of the happening in photosystem II and transported through the cycle of Quinones and released photosystem I. Electrons received by the ferrodoksin that causes the reduction of NADP become NADPH allowing cell synthesis of molecular components. On the ideal condition, the light reaction in photosynthesis reaction as follows:

$$2NADP + 3ADP + 3P + 2H_2O + 8e \rightarrow 2NADPH + 3ATP + 3P + 2H^+ + O_2$$

NADPH product is then used for the formation of glucose through reactions with the Calvin cycle.

3.2 The production of biomass and Lipid S. dimorphus

In the study, every treatment is done by harvesting as many as 50 ml sampling to quantify the dried biomass and weight the dried lipid. Uptake of lipid extraction method of Bligh-Dyer with solvent methanol, chloroform and aqua bides. Harvesting is done at the exponential phase, because in this phase of the microalgae cells grow quickly so that was expected to produce biomass. The biochemical composition of *S. dimorphus* is influenced by a variety of conditions such as nutrients, temperature, and light intensity. In the process of photosynthesis, lipids are formed from carbohydrate and protein. It is because in the third cycle of the substance met the Crab. Lipids are included in the groups of compounds rich in cluster C and H. Fats and oils is a compound that contains lipids.

The results showed that the production of biomass and lipid affected light intensity (Table 1). Based on the growth comparison, the highest biomass gain on 12,500 lux light intensity. The lowest biomass is at 10,000 lux light intensity. In accordance with the growth of the biomass, the highest number of cells grown on light intensity 12,500 lux, and the amount of the lowest cell in the 10,000 lux. The highest biomass and lipid production on the media culture with lighting 12,500 lux i.e. of 1.1407 $g.L^{-1}$ and 0.2520 $g.L^{-1}$ (22.28% dw). It is suspected that s.dimorphus can still conduct photosynthesis at 12,500 lux lighting properly, according to the highest cell densities on the growth curve. While the lowest biomass on media culture with 10,000 lux lighting of 0.6160 $g.L^{-1}$ and 0.1233 $g.L^{-1}$ (25.46% dw), according the lowest cell density and growth curve.

Table 1. The Production of Biomass and Lipid *S. dimorphus*

Sample	Production of Biomass ($g.L^{-1}$)	Production of Lipids ($g.L^{-1}$)	Productivity of Lipids ($g.L^{-1}day^{-1}$)	The Percentage of Lipids (%)
7,500 lux	0.6593^{a}	$0.1933^{a,b}$	$0.0276^{a,b}$	29.17^{a}
10,000 lux	0.6160^{a}	0.1233^{a}	0.0176^{a}	25.46^{a}
12,500 lux	1.1407^{b}	0.2520^{b}	0.0360^{b}	22.28^{a}

Description: The numbers followed the same superscript letters in one column shows not unlike real.

The biosynthesis of lipids are formed from polysaccharides that overhauled through the process of Glycolysis into Acetyl Coenzyme-A, which is prazat for the biosynthesis of fatty acids and glycerol. Fatty acid biosynthesis takes place inside the stoma, whereas the fatty acid oxidation takes place in the mitochondria. Lipid microalgae is generally in the form of glycerol and fatty acids are most numerous on the chains of C_{16} and at least there are in the chain of C_8, the fatty acids contained in the microalgae including intracellular molecular as there is in the chloroplasts [8]. Lipid *S. dimorphus* is present on the chain of C_{18} [9].

Triacylglycerol or triglycerides is lipid reserves actively synthesized in the network. The compound is used for the biosynthesis of triglycerides is L-glycerol-3-phosphate and acyl Coenzyme-A compound of fatty acids. Generally L-glycerol-3-phosphate is formed from compounds between the processes of glycolysis, namely dihidroksiaseton phosphate is converted into L-glycerol-3-phosphate by the enzyme glycerol-3-phosphate dehydrogenase with NAD^+ system attendant or NADH as a Coenzyme. The process of the acylation of the hydroxyl group from L-glycerol-3-phosphate produces lisofosfatidat acid and fosfatidat acid. The reaction is catalizied by the gliserolfosfat acyltransferase enzyme. Fatty acid acyl group was transferred from the Coenzyme A-acyl fatty acids to the hydroxyl group of the L-glycerol-3-phosphate gradually. Fosfatidat acid is hydrolyzed by the fostidat fosfatae enzyme to produce diasilgliserol. The last reaction, diasilgliserol reacts with a third molecule of Coenzyme-A acyl fatty acids are then catalyzed by diasilgliserol acyltransferase enzymes to produce triacylglycerols [10].

4. Conclusion

An increasing the intensity of light can increase cells number and lipids production *S. dimorphus*. The most optimum growth of *S. dimorphus* is on media 12,500 lux with cell number 100.8 x 10^6 $sel.ml^{-1}$ and produce biomass and lipids each 1.1407 $g.L^{-1}$ and 0.2520 $g.L^{-1}$ (22.28% dw).

References

[1] Campbell M N 2008 *Guelph Eng. J.* **1** 2-7
[2] Ahmad A L, Mat Yasin N H, Derek C J C, and Lim J K 2011. *Renewable Sustainable Energy Rev.* **15** 584-593
[3] Stone D 2003 *Biodiversity of Indonesia* (Singapore: Tien Wah Press)
[4] Khoo H H, Sharratt PN, Das P, Balasubramanian R K, Naraharisetti P K, and Shaik S 2011 *J. Bioresour. Technol.* **102** 5800-5807

[5] Yogatama R 2015 *Thesis: The Effect of Variation in Media Concentration on Biomass Production and Total Lipid of Scenedesmus dimorphus (Turpin) Kützing as Biodiesel Material* (Surakarta: Courses Of Biology, Science Faculty, Sebelas Maret University)
[6] Bligh E G and Dyer W J 1959 *Canadian Journal Biochemistry Physiology* **37** 911-917
[7] Gour R S, Chawla A, Singh H, Chauhan R S, and Kant A 2016 *PloS ONE* **11(5)** e0155321. Doi:10.1371/journal.pone.0155321
[8] Spolaore P, Joannis-Cassan C, Duran E, Isambert A 2006 *Journal of Bioscience and Bioeneginering* **101** 87-96
[9] Renaud S M V 1994 *Journal of Applies Phycology* **6** 337 – 345
[10] Wirahadikusumah M 1985 *Biochemistry: Metabolism of Energy, Carbohydrates, and Lipids* (Bandung: ITB press)

Optimization Production of Biosurfactant by *Pseudomonas putida* Using Crude Palm Oil (CPO) as Substrate

V Suryanti[1]*, D S Handayani[1], A Masykur[1] and Lindasari[1]
Department of Chemistry, Faculty of Mathematics and Natural Sciences, Sebelas Maret University, Jl. Ir. Sutami 36A Surakarta 57126 Indonesia

*Email: venty@mipa.uns.ac.id

Abstract. The production of biosurfactant by *Pseudomonas putida* has been studied. *P. putida* FNCC 0071 was grown in the nutrient broth medium supplemented with NaCl and crude palm oil (CPO). The effect of CPO concentration and fermentation time on the biosurfactant production were evaluated. The biosurfactant production was evaluated every 24 h for 10 days by optical density, surface tension and emulsification index. The best culture medium was found to be medium containing 5% v/v of CPO with 5 days of incubation time. The biosurfactant was identified as rhamnolipids.

Keywords: *biosurfactants; Pseudomonas putida; palm oil*

1. Introduction

Surfactants are surface active molecules which are usually derivatization products of petroleum oil that are non-renewable and non-degradable. Therefore, there is a high demand in replacing chemically surfactants with biosurfactants. Biosurfactants are naturally occurring amphiphilic molecules produced by a variety of microorganism. They have better environmental compatibility; e.g higher biodegradable, lower toxicity, higher selectivity and specific activity; than that of the chemical surfactants. Biosurfactants have potential application in various fields such as in food and pharmaceutics industries, oil recovery, bioremediation and biodegradation [1-4]. Nevertheless, due to their high production cost, they have not being widely used yet [5].

It is important to improve the method of biosurfactant production and characterizing the major properties of the biosurfactant. Chemical structures of biosurfactants and their physicochemical properties depends on the physical and chemical conditions under which they are grown [6]. Generally, the structure of biosurfactant relates with the structure of the substrate provided in the media and modify the substrate commonly modifies the structure and hence properties of the biosurfactants.

In recent years, research in optimizing factors for biosurfactant production has considerably increased, where numerous renewable substrates have been used, such as carbohydrates, lipids and proteins. Vegetable oils, such as soybean oil, olive oil, corn oil, castor and sunflower oil, have been used as substrates in the production of biosurfactants [7-14]. The production of sophorolipids by *Candida bombicola* have been reported using glucose and/or palm as substrates [15].

Indonesia is one of the world's biggest producer in crude palm oil (CPO) production. It is reported that 50% of produced CPO is exported in an unprocessed form, while the remaining is subject to processing into essential ingredient for food industries, washing powder, soap and biodiesel. Therefore, diversification products of CPO still need to be produced to increase the economic value of CPO. It was interesting to use CPO as a renewable substrate for the production of biosurfactants in our research.

Pseudomonas aeruginosa is considered as potentially the main producer of rhamnolipids [16]. Yet, *P. aeruginosa* species is pathogenic. Thus, our research was focused on using *Pseudomonas putida* which is generally considered non-pathogenic as producer of rhamnolipids biosurfactants. In this report, the production of biosurfactant by *P. putida* using CPO as substrate are discussed.

2. Materials and Methods

2.1. Materials

All chemical were used are analytical grade from e-Merck. The crude palm oil (CPO) was obtained from Centre for Chemical and Packaging, Jakarta, Indonesia. The strain used in this work, *P. putida* FNCC 0071, was purchased from IUC Food and Nutrition, Universitas Gadjah Mada, Indonesia.

2.2. Media Used and Growth Condition

Cultures of bacteria were maintained on nutrient agar. The media were used for optimization of biosurfactants production composed of nutrient broth (8 g/L), NaCl (5 g/L) and CPO (0, 5, 10 and 20 % v/v). Media was sterilized prior used. The flasks containing 125 mL media were inoculated with 3 ml overnight pre-culture of the strain and incubated at room temperature on a reciprocal rotary shaker at 150 rpm.

The microbial growth, surface tension and emulsification index (E24) were observed every day for 10 days of incubation. Microbial growth was evaluated by measuring the absorbance of the culture at 600 nm. The capillary rise method was used for measuring the surface tension. The surface tension was calculated using equation (1), where γ_a is aquadest surface tension (72.13 mN/m), γ_x is surface tension of x (dyne/cm), h_a is aquadest height, h_x is height of x, d_a is aquadest density (0.997 gr/cm^3) and d_x is density of x (gr/cm^3).

$$\gamma_x = \frac{h_x \rho_x}{h_a \rho_a} \gamma_a \quad \text{.......................................} (1)$$

The E24 was determined by adding 1 mL of vegetable oil to the same amount of aqueous culture containing biosurfactants, mixing with a vortex for 2 minutes, and leaving to stand for 24 h. The E24 index is given as the height of the emulsion layer, divided by the total height, multiplied by 100.

2.3. Biosurfactants isolation and FT-IR analysis

Culture liquid of *P. putida* was filtered by Buchner filtrations using Whatman filtration paper grade 42. The supernatant was then acidified to pH 2.0 with HCl 6 N and leaved overnight at 4 °C. The supernatant was extracted twice with *n*-hexane for removing of the remaining CPO. The aqueous solution was then extracted twice using ethyl acetate. Sodium sulphate anhydrous was then added to the organic layer. The organic layer which was free form water was then evaporated. The sample was then analysed by FT-IR Spectroscopy. The spectra of the sample was recorded at room temperature as KBr pellets on Shimadzu type FT-IR 8201 PC spectrophotometer. The light source of transmittance was in the range of 4000-500 cm^{-1}.

3. Results and Discussion

The effect of addition of CPO and fermentation time on the biosurfactant production were examined. Incubation was carried for 10 days where the optical density, surface tension and E24 were noted every day (Fig. 1). Biosurfactant production optimization was carried out in media containing variation amount of CPO (5, 10 and 20 %). The media with no addition of CPO was also performed as a negative control. The optimum condition for the biosurfactant production was explored for the highest emulsification index and the lowest surface tension.

Figure 1(a) presents growth optimization of biosurfactants production using CPO as substrate. The results showed that the addition of CPO in all concentration enhanced the growth of *P. putida* culture significantly. The highest emulsification index was achieved for the media containing 5% v/v of CPO on 5 days of fermentation (Fig. 1(b)). It is noted that the lowest surface tension was obtained for the media containing 5, 10 and 20% v/v of CPO on 5 days of fermentation. These three media have similar value of surface tension (Fig. 1(c)). Base on the Duncan statistical analysis for the emulsification index and surface tension data, the media containing 5% v/v of CPO on 5 days of fermentation was selected as the optimum condition for the biosurfactants production.

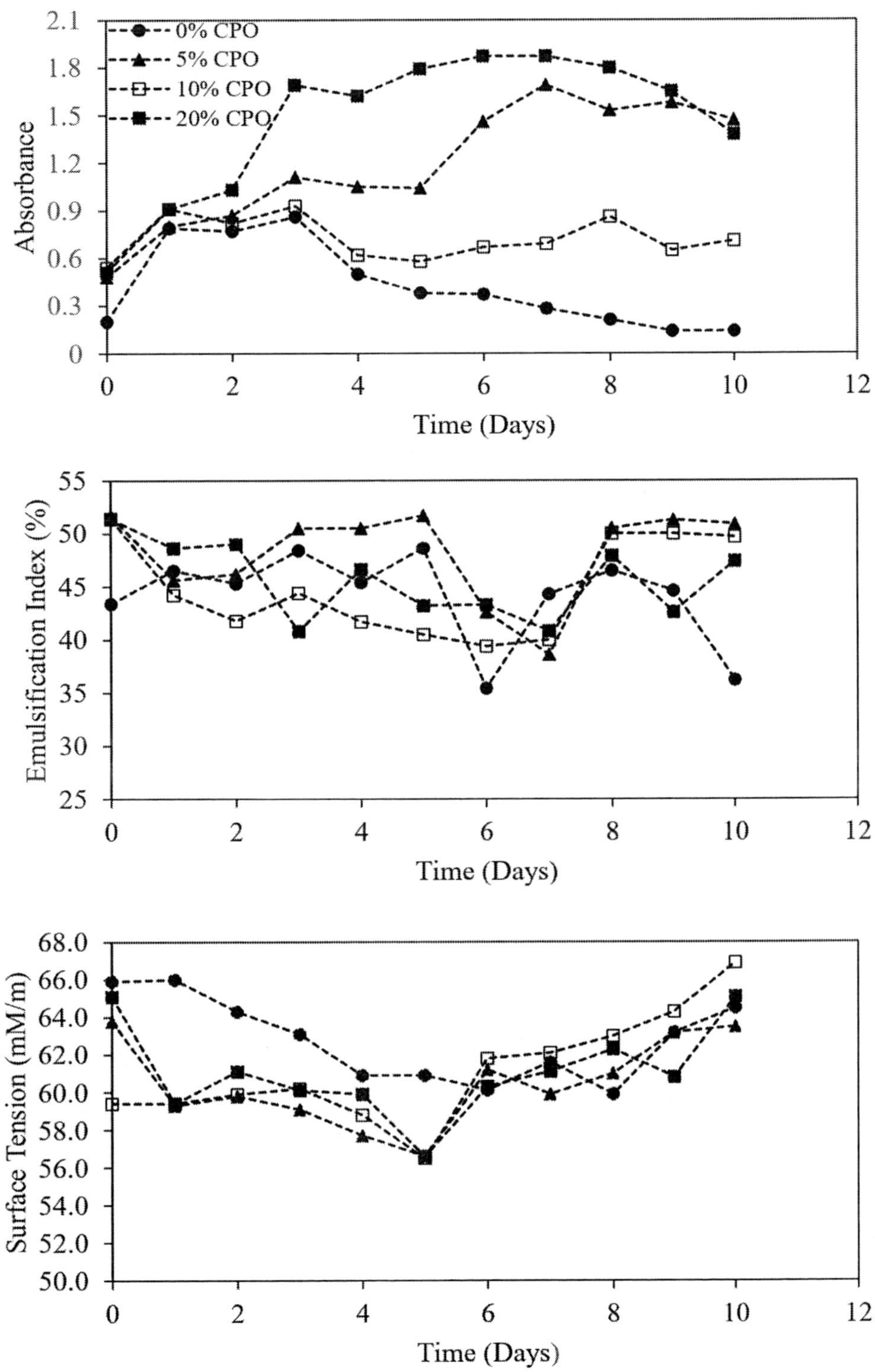

Figure 1. Biosurfactants production by *P. putida* grown on CPO was monitored for: (a) optical density, (b) emulsification index and (c). surface tension.

The production of the biosurfactants was applied for 125 mL media containing 5% v/v of CPO in 250 mL flask for 5 days of fermentation and resulted 0.503 g of biosurfactants for 250 mL media. The FT-IR analysis was performed to identify the functional groups present in the biosurfactants (Fig. 2). The O-H stretching vibration has been observed by a strong and broad band at 3316 cm^{-1} due. The band at 2955 and 2925 cm^{-1} confirmed the presence of the –CH stretch of methylene. The asymmetric stretching vibration of C=O was characterised by the band at 1740 cm^{-1} for the carboxylate and 1660 cm^{-1} for the ketone. The presence of alkyl chains in the compounds was confirmed by the band at 1433 as a result of deformation and bending vibrations of aliphatic chains (-CH_3- and CH_2-). The stretching of C-O ester was suggested by the presence of the intense and strong band at 1166 cm^{-1}. The rhamnolipids biosurfactants were produced by *P. putida* which was confirmed from the obtained peaks of biosurfactants shown in FT-IR spectrum (Fig. 2). This results was supported by earlier studies reported by other researchers that glycolipids usually produced by *Pseudomonas* spp [17-20].

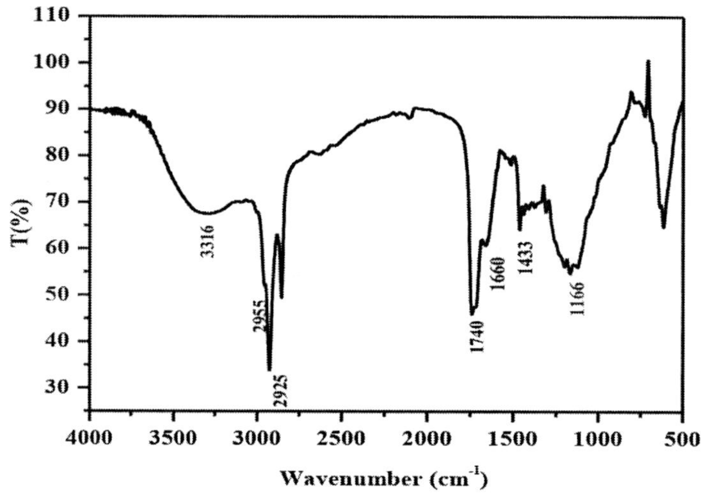

Figure 2. FT-IR spectrum of the produced biosurfactant.

4. Conclusions

Biosurfactant production by *P. putida* in crude palm oil was developed, where the effect of medium composition and fermentation time were evaluated. The optimum condition for biosurfactant production was achieved when a medium containing of 8 g/L nutrient broth, 5 g/L NaCl and 5% v/v crude palm oil was used as media with 5 days of fermentation. Biosurfactant was identified as glycolipids type biosurfactant, which was rhamnolipid.

Acknowledgments

The authors acknowledge their thanks to Ministry of Research, Technology and Higher Education Indonesia for the financial support to this research in the form of Insentif Riset Sistem Inovasi Nasional (Insinas) 2016 with contract number: 453/UN27.21/LT/2016.

References

[1] Ron E Z and Rosenberg E 2001 *Environ. Microbiol.* **3** 229-236.
[2] Muthusamy K, Gopalakrishnan S, Ravi T K and Sivachidambaram P 2008 *Curr. Sci.* **94** 736-774.
[3] Banat I M, Franzetti A, Gandolfi I, Bestetti G, Martinotti M G, Fracchia L, Smyth T J and Marchant R 2010 *Appl. Microbiol. Biotechnol.* **87** **2** 427-444.

[4] Pacwa-Płociniczak M, Płaza G A, Piotrowska-Seget Z and Cameotra S S 2011 *Int. J. Mol. Sci.* **12** 633-65.
[5] Banat I M, Makkar R S and Cameotra S S 2000 *Appl. Microbiol. Biotechnol.* **53** **5** 495-508.
[6] Ghazali R and Ahmad S 1997 Biosurfactans-A Review *Elaeis* **9** **1** 34-54.
[7] Makkar R S and Cameotra S S 2002 *Appl. Microbiol. Biotechnol.* **58** 428-434.
[8] Ferraz C, De Araujo A A and Postore G M 2002 *Appl. Biochem. Biotechnol.* **98-100** 841-847.
[9] Pantazaki A A, Dimopoulou M I, Simou O M and Pritsa A A 2010. *Appl. Microbiol. Biotechnol.* **88** 939-951.
[10] Accorsini F R, Mutton M J R, Lemos M G M, Benincasa M 2002 *Braz. J. Microbiol.* **43** **1** 116-125
[11] Sadouk Z, Hacene H and Tazerouti A 2008 *Oil Gas Sci. Technol. – Rev. IFP* **63** **6** 747-753
[12] Suryanti V, Hastuti S, Wahyuningsih T W, Mudasir and Muliawati D I 2009 *Indo. J. Chem.* **9** **1** 107-112
[13] Suryanti V, Marliyana S D and Andriani D 2016 *IOP Conf. Ser. Mater. Sci. Eng.* **107** 012018 doi:10.1088/1757-899X/107/1/012018.
[14] Suryanti V, Hastuti S and Pujiastuti D 2016 *AIP Conf. Proc.* **1710** 030016 doi: 10.1063/1.4941482.
[15] Sasidharan V, Thambirajah J J, Ho, C C and Hasyim M A 1993 *The 16th Malaysian Microbiology Symposium* 6264.
[16] Abdel-Mawgoud A M, Lépine F and Déziel E 2010 *Appl. Microbiol. Biotechnol.* **86** **5** 1323-1336.
[17] Lee Y, Lee S Y and Yang J-W 1999 *Biosci. Biotechnol. Biochem.* **63** **5** 946-947.
[18] Santa Anna L M, Sebastian G V, Menezes E P, Alves T L M, Santos A S, Pereira Jr N and Freire D M G 2002 *Braz. J. Chem. Eng.* **19** **2** 159-166.
[19] Rashedi H, Jamshidi E, Assadi, M M and Bonakdarpour B 2005 *Int. J. Environ. Sci. Tech.* **2** **2** 121-127
[20] Onbasli D and Aslim B 2009 *J. Environ Biol.* **30** **1** 161-163.

Bioethanol Production By Utilizing Cassava Peels Waste Through Enzymatic And Microbiological Hydrolysis

R G Witantri[1], T Purwoko[1], Sunarto[1], E Mahajoeno[1]
[1] Biology, Faculty of Mathematics and Natural Sciences, Universitas Sebelas Maret, Surakarta

E-mail: rekyangaluhwitantri @gmail.com

Abstract. Cassava peels waste contains, cellulose which is quite high at 43.626%, this is a potential candidate as a raw for bioethanol production. The purpose of this study was to determine the performance of the enzymatic hydrolysis, microbiological (Effective microbe) and fermentation in cassava peel waste is known from the results of quantitative measurement of multiple ethanol parameters (DNS Test, pH, ethanol concentration). This research was carried out in stages, the first stage is hydrolysis with completely randomized design with single factor variation of the catalyst, consisting of three levels ie cellulase enzymes, multienzyme and effective microbial EM4. The second stage is fermentation with factorial randomized block design, consisting of three groups of variations of catalyst, and has two factors: variations of fermipan levels 1, 2, 3% and the duration of fermentation, 2,4,6 days. The parameters in the test is a reducing sugar, pH and concentration of ethanol. The results showed that variation of hydrolysis treatment, fermentation time, and fermipan levels has real effect on the fermentation process. On average the highest ethanol content obtained from the treatment with multienzyme addition, with the addition of 2% fermipan levels and on the 2nd day of fermentation that is equal to 3.76%.

1. Introduction

One of solutions for the energy crisis is the utilization of bioethanol as alternative energy. Bioethanol can improve combustion efficiency because it contains 35% oxygen beside that, bioethanol is eco-friendly fuel because not produce carbon gas pollution[1]. In this era, bioethanol produced from staple food raw materials such as molasses from sugar cane, corn, and cassava tubers. Waste from industrial agriculture are potentially used as raw material for bioethanol because of its availability is still abundant, the cost is relatively cheap and could reduce competition with staple food[2]. Cassava production in Indonesia reach 22 million tons in 2015. From the cassava processing, cassava peel waste which generated is high and unused optimally. Utilization of cassava peels is only for animal feed or simply disposed off as trash. Lignocellulose material of cassava peels can be used as raw material for bioethanol[3]. Cassava peels waste contains 43.626% cellulose, hemicellulose and lignin 10.384% 7.646%[1]. But in Indonesia, the production of bioethanol is constrained by the high cost of production budget both in industrial scale and cottage industry scale, it causing the conversion of conventional energy to renewable energy can not be applied optimally in Indonesia.

There is two main stages of bioethanol production, first is hydrolysis process that break down cellulose into simple sugars with the help of acids or enzyme. Second is fermentation process that will convert sugar into alcohol through anaerobic respiration by microbes[4]. In this study, enzymes is used as catalysts for the break down reaction of lignocellulose into simple sugars . Enzymatic hydrolysis is relatively more efficient in producing sugar because it can minimize the formation of toxic compounds as a result of acid hydrolysis process. However, the disadvantage of these methods is it takes much longer time than acid hydrolysis process and high cost of purified enzyme material[5]. To overcome the high cost of using purified enzyme, in this study multienzyme and EM4 (effective microbe) is used, to obtain multienzyme and EM4 performance the purified enzyme (cellulase) is used as positive comparison.

Multienzyme and effective microbial EM4 commercially known as an additional supplements for livestock. Addition of multienzyme and effective microbial EM4 on animal feed, can improve the digestibility especially for foods that contain high fiber cellulose so the nutrition could be optimally absorbed[6-7]. Based on the composition of cassava peel as lignocellulosic biomass, multienzyme has appropriate enzyme content, which can be used on enzymatic hydrolysis process such as Galactosidase, xylanase, Galaktomananase, β-glucosidase, amylase. Galactosidase enzyme will catalyze the hydrolysis of bonds α-ᴅ-galaktosidik contained in the hemicellulose galakto- oligosaccharides[8]. EM4 (Effective Microorganisms) consists of three main components, *Rhodopseumonas palustris*, *Lactobacillus casei*, and *Saccharomyces cerevisiae*. *Rhodopseumonas palustris* prove has capability to degrading aromatic compounds such as coumaroyl amide, feruoyl amide, coumaric acid and benzoic acid contained in lignin through benzoyl-CoA pathway. The degrading of lignocellulose matter will produce aromatic compunds that will inhibit hydrolysis and fermentation process. *Rhodopseumonas palustris* could degrade aromatic compounds anaerobically without damaging or consuming glucose and xylose which is needed during the fermentation process[9]. Both materials are affordable and can be obtained easily by the general public. The main objective of this study was to evaluate the usability of agricultural waste, especially cassava peel waste for bioethanol raw material, using various treatment on hidrolysis process such as: multienzyme, effective microbes in the EM4, and cellulase enzymes.

2. Experimental

2.1 Material preparation

Cassava peels are dried in the sun for 3 days. It is dried to remove moisture and to obtain dry biomass. To simplify the process of hydrolysis, cassava peel waste crushed using a blender. In the first step is hydrolysis with 3 different treatments, and 3 replications. To make 1 sample a ratio of 1: 4 is used , 25 grams of cassava peel waste were given 100 mL of distilled water.

2.2 Hydrolisis

Hydrolysis processes aimed to convert cellulose into glucose[5]. To catalyze the reaction, enzymes and effective microbes at EM4 is used. Each liquid sample was treated differently by adding a cellulase enzyme (Ka), multienzyme (Kb), and effective microbial EM4 (Kc). Control treatment (K0) without the addition of enzymes or other substances. To make Ka samples, 2.5 mg of cellulase enzymes were mixed into the liquid sample. Kb sample was prepared by adding 5 g multienzyme into liquid samples, two types of samples are conditioned in the range of 5- pH 5.5 using hydrochloric acid, or HCl 0.1N. Meanwhile, to make Kc sample, liquid samples were given 5 ml EM4 by conditioning the pH of 6-6.5. For a sample K0 pH is conditioned in the range of 5-5.5. Hydrolysis process is carried out at room temperature. Conditioning of pH and temperature aimed to optimize the performance of the enzyme. Hydrolysis process is carried out for 24 hours. After the hydrolysis process is finished, next step is the samples were sterilized by autoclaving at 121°C temperature for 15 minutes. Sterilization aims to stop the activities of enzymes[5]. DNS method is used to determine reducing sugar which is after the hydrolysis process.

2.3 Fermentation and distillation

The second stage is fermentation, samples which has been hydrolyzed then given inoculant in the form of fermipan that contain yeast *Saccharomyces cerevisiae*. The fermentation process aims to transform sugars into ethanol through a process of anaerobic respiration by microbiological agent[4]. Fermipan will be varied to 3 levels content ie 1%, 2% and 3% of the sample weight. The fermentation process was conducted for 6 days in room temperature conditions. Measurement of pH and reducing sugar measurements performed on day 2, day 4 and day 6 during fermentation. Ethanol content analysis performed on day 2, day 4 and day 6 during fermentation. During the fermentation process, the sample bottle is connected by a rubber hose and the hose is inserted into another bottle filled with water to minimize direct contact with air. After fermentation is complete sample were filtered and then distilled. Solution is poured into the distillation flask, heater temperature is maintained at a temperature of 70°C. The distillation process is done for 15-30 minutes. Distillate that has been obtained then measured the density of ethanol[5].

2.4. Analysis of reducing sugar content

Reducing sugar measurements using DNS method refers to the theory of Miller (1959)[10]. DNS reagent 1% was prepared by dissolving 5 g DNS, 1 g of phenol, 0.25 g Na-metabisulfite, and 5 g NaOH in 300 mL of distilled water. The solution is then diluted to 500 mL using distilled water and then filtered after 24 hours. In addition to DNS solution, rochelle salt (KNA-tartrate) 40% is produced with dissolving 40 g KNA-tartrate into 100 mL of distilled water.

A total of 1 mL sample were taken and put into a test tube and was given 2 mL of 1% DNS reagent, the mixture is then alloyed using vortex. Furthermore, test tubes containing samples and DNS 1% put in boiling water for 5 minutes to allow the reduction process of reducing sugar DNS to run faster and optimized, further 1 mL KNA-tartrate 40% was given, shortly before the solution start to being cool. KNA-tartrate 40% together with a solution of Na-metabisulfite is used to remove dissolved oxygen glucose so it can protect from damage caused by dissolved oxygen. Absorbance measurements using UV-VIS spectrophotometer with wavelength 540 nm.

2.5. Analysis of ethanol content with density analysis

Bioethanol content analysis using pycnometer density method at temperature of 20°C. To determine the concentration of ethanol, picnometer volume and density of ethanol is calculated using the following formula:

1. Calculate the volume of pycnometer

$$c : \frac{b - a}{\rho \text{ water}} \quad (1)$$

2. Calculating Ethanol Density

$$\text{Density} : \frac{d - a}{c} \quad (2)$$

Information
a: weight of the empty pycnometer
b: weight of pycnometer containing distilled water (g)
c: Volume Pycnometer (mL)
d: Weight Pycnometer content of solution (g)
ρ water: 0.998203 g / mL (density of water at a temperature of 20°C)

From the density number, ethanol percentage could determined by compare it with standard density of ethanol in AOAC table (2005).

3. Result and Discussion

3.1. Reducing sugar content on enzymatic hydrolysis

Table 1. Reducing Sugar Content that Resulted from Enzymatic Hydrolisis g/100 ml

Treatment	Reducing Sugar Content (g/100 ml)
Cellulose	8.0667
EM4	8.7900
Multienzyme	11.0267
Control	4.700

From Table 1 it is known that the highest levels of reducing sugars produced by the multienzyme treatment, while EM4 treatment can produce the highest reduction sugar after multienzyme i.e. 8.7900 g/L. Between reducing sugar produced by EM4 and reducing sugar produced by cellulases do not show much difference. Cellulase can produce high levels of reducing sugars of 8.0667 g/L, while control produces a low reducing sugar ie 4.770 g/L. Reducing sugar greatly affect the levels of ethanol formed, the higher the sugar content reduction of the ethanol content will also be higher.

Multienzyme has the highest reducing sugar because multienzyme consists of various types of complex enzymes that are able to do the optimumconversion process from lignocellulose into sugar, other than that there is also an amylase enzyme that can hydrolyze the starch content cassava in the skin so the result is more optimal. Cellulase enzymes have lower glucose levels when compared to multienzyme ie 8.0667 g/L, it is because of cellulase enzymes only can hydrolyze cellulose content found on the peels of cassava. Cellulase is a complex mixture of several enzymes that work together synergistically to attack parts of the distinctive cellulose fibers[11].

EM4 has a reducing sugar which is lower than the multienzyme that is equal to 8.7900 g/L. This is because EM4 a livestock supplemen, containing effective microbes in contrast with the multienzyme consisting of different types of enzymes. When added into sample, the microbes originating from EM4 takes time for the adaptation phase until it can produce the enzyme in the exponential phase. Can be interpreted within a period of 24 hours in the process of hydrolysis, microbial EM4 not work optimally in producing sugar reduction. In the control treatment reducing sugar produced the lowest compared to the three other treatments, this indicates multienzyme, EM4 or cellulase has an influence in increasing the reducing sugar hydrolysis process on cassava peel waste. In the control treatment, the hydrolysis process still produces reducing sugar because in cassava peel contained the endogenous enzymes. The endogenous enzymes such as β-glucosidase enzyme, known as linamarase. Linamarase will hydrolyze glucoside compound linamarin and lotaustralin cyanogenik form spontaneously when the damage occurred mechanically in a network or as a result of microbial activity.

3.2. Reducing sugar recidue on during fermentation process

In producing ethanol at anaerobic condition, *Saccharomyces cerevisiae* will consume glucose as the main carbon source[12]. Based on Figure 1, reducing sugars is drop constantly fom the day of two fermentation until day of six fermentation. According to the several result on bioethanol research, reduction of reducing sugar during fermentation process, because the sugars in the media used by *Saccharomyces cerevisiae* as a carbon source for ethanol synthesis[13]. Reducing sugar which is decreased may indicate ethanol production, but according to Bulawayo (1996) in Meldha (2012)[14] revealed that the decrease of reducing sugar that is not accompanied with elevated levels of bioethanol, which is produced due to the sugar consumed is used only for the growth of yeast, so not all sugars can be converted into ethanol.

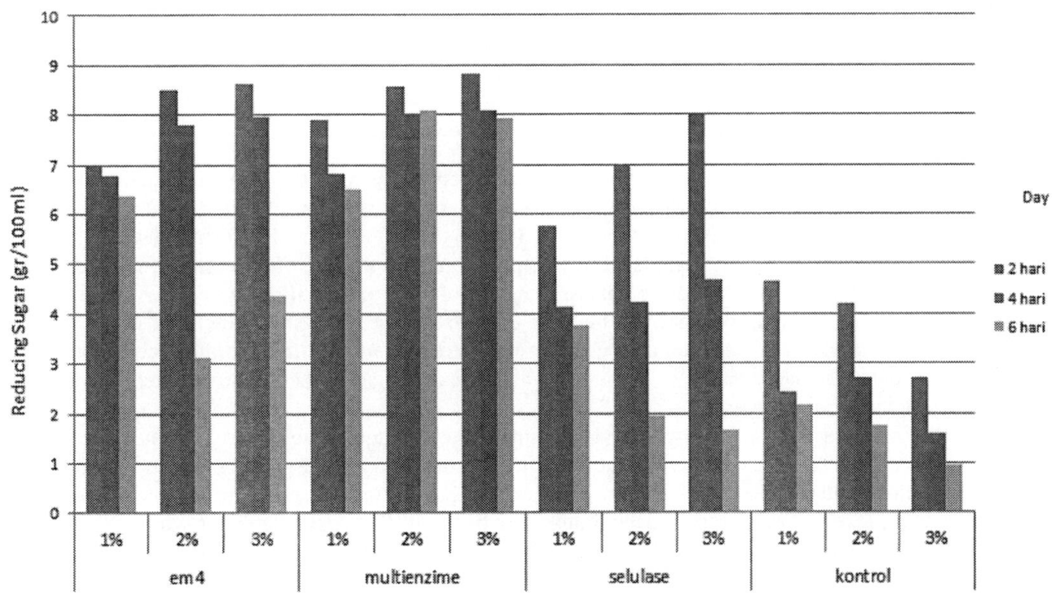

Figure 1. Reducing Sugar Fluctuations during Fermentation Process

Based on the variation of fermipan content level shows that the average usage levels of reducing sugars in fermipan 2% had the highest amount ie 1,6566 g/L, while the fermipan 3% on average ie 1,6733 g/L, and the lowest one is from fermipan 1% i.e. 0.8170 g/L (Table 2). However fermipan content level comparable to the number of cells contained in the substrate, so that the higher the number of cells, the use of a reducing sugar will also high, because the yeast have a rapid growth phase[15].

IOP Conf. Series: Earth and Environmental Science 75 (2017) 012014 doi:10.1088/1755-1315/75/1/012014

When the cell density reaches a maximum then it will trigger a stationary phase, which means the number of living cells is equal to the number of died cells, then it will followed by death phase, in this phase the amount of yeast that dies will increase until it will eventually happen death cell in bulk[12]. In contrast the number of cells that are too low will result in the reduction of sugar consumption by running slowly making yeast fermentation process is also slow.

Table 2. The Average of Reducing Sugar Utilized by Yeast

	Fermipan Level		
	1%	2%	3%
	0.2304	0.7171	0.6530
	0.4192	4.6561	3.6048
	1.0927	0.5808	0.7295
	0.3193	-0.0953	0.1556
	1.6097	2.7553	3.3472
	0.3915	2.3102	3.0287
	2.2375	1.5211	1.1104
	0.2359	0.9411	0.6237
Average	0.8170	1.6733	1.6566

3.3 Ethanol yield content

Table 3. Ethanol Yield Percentage (%)

Day of Fermentation	EM4			Multienzyme			cellulose			Control		
	1%	2%	3%	1%	2%	3%	1%	2%	3%	1%	2%	3%
2 day	2.50	3.15	3.27	3.14	3.76	3.66	2.38	2.63	3.17	1.72	2.00	1.70
4 day	2.46	2.89	3.30	2.63	3.24	3.42	2.43	2.45	2.61	2.38	2.57	2.34
6 day	2.48	2.22	2.68	2.73	3.21	3.45	2.09	2.22	2.50	1.81	2.23	1.90

As a result of primary metabolism, ethanol is produced along with the growth of cells. In fermipan content level variation, the use of fermipan 2% or 3% could increase ethanol production rather than the use of fermipan 1%. Tthehe tabtable le le 3 show that ethanol yield percentage with 2% and 3% fermipan is higher rather than fermipan 1%.This point out that the rise of fermipan content level on the substrate it will followed by the increasing of ethanol rate production by yeast. According to the research by Zely (2014)[16], that the number of yeast cells affect the content of ethanol formed. Increasing the number of yeast on the substrate will accelerate the fermentation process, especially if the sugar content in the substrate is high. Yeast activities in sugar fermentation will increase so that the ethanol content could increased.

In the three variation of the hydrolysis treatment the average value of the highest ethanol content generated on the 2nd day of fermentation, after fermentation time passing day of 2, the ethanol concentration tends to decrease as shown in Table 3. It is directly proportional to the decrease of reducing sugar levels at the time fermentation, the possibility is reducing sugar which is used by the yeast is not converted to ethanol. The decrease in reducing sugar levels are not accompanied by the increase of ethanol content because the sugar used by yeast for cell growth or survival without can be converted into ethanol. On the 2nd day, ethanol reach the highest percentagr because yeast cells are in the exponential phase, in this phase the yeast cells will produce ethanol and growth rapidly. After the 2nd day of fermentation, the increasing of cells density number because of rapid growth trigger cells to enter stationary phase cells so that the cells no longer produce ethanol makes the optimum concentration of ethanol drop slowly. The level of ethanol production is influenced by the activity of the cells, whereas cell activity is influenced by the type of media, and environmental conditions (eg temperature and pH).

The different result show from control treatment, ethanol yield percentage increase in the day of four fermentation and then drop at sixth day it is because the slow consumption of reducing sugar by yeast or different type of sugar which is produce from hydrolysis process with control treatment. Cassava peels wastes contain cyanogenic glucose named linamarin and lotaustralin that will be degraded by endogenous enzyme linamarase (β-glucosidase enzyme) .

Based on the average value of ethanol content in Table 3, the highest ethanol content is generated by multienzyme ie 3.76% using 2% fermipan. In EM4, the highest ethanol content produced from by fermipan 3% ie 3.27% and the cellulase highest levels of ethanol produced at the levels of use fermipan 3% at 3.17%, control treatment could produce 2,57% ethanol at fourth day of fermentation using fermipan 2%. These results are appropriate with reducing sugar obtained from hydrolysis process presented in Table 7, the highest reducing sugar produced by multienzyme amounted to 11.0267 g/L. The highest reducing sugar content will produces high percentage of ethanol yield.

4. Conclusion

The highest reducing sugar which produce from enzymatic hydrolisis is deliver from Multienzyme treatment (11,0267 g/100ml) it proves that multienzyme has a better performance compare to purified enzyme cellulase. The highest of ethanol yield precentage that is produce from fermentation process is 3,76 % with the addition of 2% fermipan level; on the second day fermentation using multienzyme treatment.

References

[1] Artiyani A 2011 Bioetanol dari Limbah Kulit Singkong melalui Proses Hidrolisis dan Fermentasi dengan *Saccharomyces cerevisiae* Thesis (Surabaya: Program Magister Jurusan Teknik Lingkungan Institut Teknologi Sepuluh Nopember)
[2] Sarkar N, Ghosh S K, Bannerjee S, Aikat K 2012 *Renewable Energy* **37** 19
[3] Hikmiyati N, Dan N S, Yanie 2008 *Jurnal Teknik Kimia Universitas Diponegoro* **1**1
[4] Fachry A R, Astuti P and Puspitasari T G 2013 *Jurnal Teknik Kimia* **19** 60
[5] Novia N, Faizal M, Ariko M F and Yogamina D H 2011 *Prosiding Seminar Nasional AvoEr* **3** 451-462
[6] Djaya M S and Hidayat M I 2013 *Polhasains* **1**1
[7] Suciani K P W, Sumardani N L G, Bidura I G N G, Kayana I G N, and Lindawati S A 2011 *Jurnal Veteriner Indonesia* **12** 69
[8] Fernandez M B, Gulshan A 2014 *Bioethanol and Recombinant Proteins Production from Milk Whey and Molasses* (Italy, University of Coruna)
[9] Austin S, Kontur W S, Ulbrich A, Oshlag J Z, Zhang W, Higbee A, Zhang Y, Coon J J, Hodge D B, Donohue T J 2015 *Environ. Sci. Tech.* **49** 8914
[10] Miller G L 1959 *Anal.l Chem.* *3* 426
[11] Hamelinck C N, Hooijdonk GV, Faaij A P C 2005 *Biomass Bioenergy* **28** 384
[12] Azizah N, Al-Baarri A N, S Mulyani 2012 *Jurnal Aplikasi Teknologi Pangan* **1** 72
[13] Putri L S E and Sukandar D 2008 *Biodiversitas* **9** 112
[14] Meldha Z, Chairul, Amraini S Z 2012 *Research Gate* **1** 1
[15] Yumas M and Rosniati D 2014 *Biopropal Industri* **5** 13
[16] Zely D F 2014 *Pembuatan Bioetanol dari Serabut Kelapa* Unergraduate Thesis (Bengkulu: Jurusan Biologi Universitas Bengkulu)

IOP Conf. Series: Earth and Environmental Science **75** (2017) 012015 doi:10.1088/1755-1315/75/1/012015

Synthetic of Allophane-Like and Adsorption for Lead (Pb) Heavy Metals

Pranoto[1], C Purnawan[1] and A D Widyasari[1]

[1] Chemistry Department, Faculty of Mathematic and Natural Science, Sebelas Maret University

Abstract. A synthesis allophone-like had been done from aluminum nitrate nonahidrat (Al $(NO_3)_3.9H_2O$) and tetraethyl orthosilicate (TEOS). This study aims to determine synthesis of allophane-like based on variations in the ratio of Al/Si is applied as an adsorbent for the lead (Pb) metal. The study was conducted with ratio variation Al/Si 0.5; 0.75; 1; 1.25 and 1.5 from solution volume nonahidrat (Al $(NO_3)_3.9H_2O$) and TEOS at pH 3-4 calcined at temperature of 400°C. The adsorption type is determined by the adsorption isotherm. The results of the research shows that allophane-like can be synthetized from a solution of nonahidrat (Al $(NO_3)_3.9H_2O$) and TEOS with a variation of the ratio of Al/Si 0.5; 0.75; 1; 1.25 and 1.5. Allophane-like has function groups of -OH group; spanning asymmetrical form of O-Si-O/O-Al-O; Si-O/Al-O bending vibration and relatively weak absorption which strengthen their OH group. The results of allophane-like adsorption capacity of 1.61 mg/g in the ratio Al/Si of 1.5 at pH 4 with a contact time of 30 minutes with a percentage of 80.77% adsorption. The adsorptions of Pb heavy metals with allophane-like tend to follow the type of adsorption isotherms of Freundlich or physics.

1. Introduction

Indonesia is one of the most active volcanic regions in the world that has abundant resources of Andisol soil. Andisol soil contained mineral allophane alumino silicate hydrate with a chemical composition that is not fixed and is shown in a ratio Al/Si which can vary 1-2. Allophane has diameter of 3-5 nm with high surface area and active function groups named silanol (Si-OH) and aluminol (Al-OH). Allophane has absorption and high cation exchange so that it can be used as an adsorbent, for example, the adsorbent of heavy metal ions in industrial waste [1].

In the previous research, allophane-like synthesis has been done using the ratio of Al/Si, nonahidrat solution of aluminum nitrate nonahidrat (Al $(NO_3)_3.9H_2O$) and tetraethyl orthosilicate (TEOS) with a ratio Al/Si 2; 1.5; 1 and 0.5 [2]. According to Denaix L *et al.* (1999) [3], allophane with a ratio Al/Si = 2 has similarities with imogolit with smaller pore diameter which is 2.4 nm. The Allophane heated at a temperature of 400°C is able to open the pores so it will increase the power to adsorp [4]. Therefore, the study was conducted synthetic allophane-like by using a solution of TEOS and nonahidrat (Al $(NO_3)_3.9H_2O$) with the ratio Al/Si 0.5; 0.75; 1; 1.25 and 1.5 and then heated at a temperature of 400°C. Allophane-like can be used as an alternative of heavy metal adsorbent. Heavy metals are used in this study is a lead (Pb) heavy metal. Dispersion element Pb can also occur due to the disposal of waste from metal mining industry. This should be wary because it could pollute the environment that may lead to the emergence of various illness or death.

2. Experimental

2.1. The Synthetic and Characteristics of allophane-like

The conduction of allophane-like was started by adding TEOS solution to nonahydrate (Al $(NO_3)_3.9H_2O$) 0.1 M solution. The ratio Al/Si was made according to the Table 1.

Table 1. Ratio Al/Si allophane-like

Ratio volume Al/Si	Ratio mol Al/Si	Volume $Al(NO_3)_3.9H_2O$	Volume TEOS
0.5	0.006	50 mL	100 mL
0.75	0.009	50 mL	66.67 mL
1	0.012	50 mL	50 mL
1.25	0.015	50 mL	40 mL
1.5	0.018	50 mL	33.33 mL

150 mL NaOH solution put into the sample solution with the addition of 50 mL/hr with stirring and heating at room temperature. The pH of the solution set 3-4 with the addition of concentrated H_2SO_4 in a state of agitation. The sample solution is kept overnight to form a gel. The gel was heated for 7 days at a temperature of 100°C [2]. The samples were dried calcined at a temperature of 400°C.

2.2. Characterization of allophane-like

In the characterization of allophane-like the activity process were done with function group analysis of allophane-like, crystalinity analysis of allophane-like, morphology analysis of allophane-like, surface analysis of allophane-like, acidity analysis of allophane-like.

2.3. Adsorption Capacity of allophane-like to Pb metal

In the adsorption capacity of allophane-like for Pb ions metal were conducted solution of HNO_3 0.05 M, and standard solution of Pb ions of 0 ppm; 1 ppm; 2 ppm; 3 ppm and 4 ppm. The determination of adsorption isotherms was done in optimum condition.

3. Result and Discussion

Allophane was synthesized from TEOS as silica source solution and nonahidrat (Al $(NO_3)_3.9H_2O$) as a source of alumina. The ratio Al/Si used was 0.5; 0.75; 1; 1.25 and 1.5. The ratio of Al/Si was made using a volume ratio of nonahidrat (Al $(NO_3)_3.9H_2O$) with volume variation of TEOS solution. The material made by using the ratio of the volume will produce a different mole ratio with natural allophane so that the material is called allophane-like or a material which has similarities with allophane. Allophane-like has shape and white turbid.

3.1. Analysis of functional groups allophane-like

IR analysis results for each ratio Al/Si allophane-like shown in Figure 1.

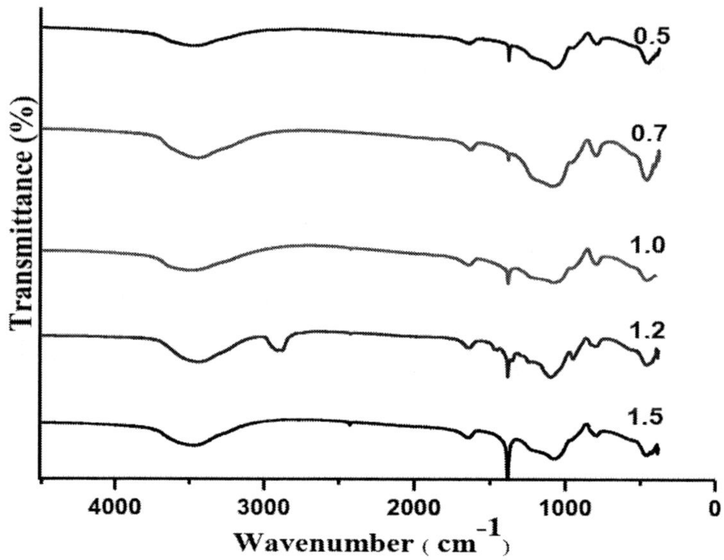

Figure 1. IR spectrum for allophane-like ratio Al/Si = 0.5; Al/Si 0.75; Al/Si = 1; Al/Si = 1.25 and Al/Si = 1.5

Based on the results of FTIR analysis, each ratio of Al/Si has four main areas of infrared absorption. Functional areas shown in the wave number 3439; 3443; 3484; 3448 and 3463 cm^{-1} which is -OH of H_2O. The wave number about 1600 cm^{-1} 1 indicates strengthen the absorption of H-OH group. Regional wave number of 962-1247 cm^{-1} indicates the group O-Si-O or O-Al-O. Regional wave numbers 408-563 cm^{-1} showed absorption of Si-O or Al-O. In the area of wave number of about 1400 cm^{-1} which appears on each ratio Al/Si shows the area of Al-OH or Si-OH is often called the fingerprint region [5].

3.2. Analysis crystallinity of Allophane-like

In diffractogram of allophane-like do not show high crystallinity of allophane-like at all ratios Al/Si. The allophane-like sample showed an amorphous structure with a sloping peaks. According to [6] diffractogram allophane have amorphous form. Allophane-like has a low crystallinity because the calcination temperature treatment performed on samples at low temperatures [7]. The heating process at allophane-like performed at a temperature of 400°C construct amorphous structure with high irregularity of crystalline allophane.

In this study, it is also conducted the heating at a temperature of 600°C. The results of diffractogram of allophane-like at the heating temperature of 600°C has a crystallinity higher than the heating temperature of 400°C. Diffractogram of allophane-like with a heating temperature of 600°C has similarities with allophane diffractogram according to the data JCPDS number 38-0449. Allophane-like at the heating temperature of 600°C is not used as an adsorbent to heavy metals Pb due to have a small adsorption capacity as an adsorbent. According to Meima *et al.* (2002) [8] , allophane have no a high thermal resistance so that if it is done in a high-temperature heating, the surface area will decrease and cause the adsorption capacity decreases. According to Reinert *et al.* (2011) [4], the maximum adsorption capacity allophane-like performed at a temperature of 400°C. Allophane-like has a lower crystallinity compared with natural allophane. Natural allophane still contains several other minerals such as feldspar, gibbsit, kaolinite and montmorillonite.

3.3. Analysis of the surface morphology of the allophane-like

The results of SEM analysis can be seen that the allophane-like ratio Al/Si 0.5; 0.75; 1; 1.25 and 1.5 have morphology chunks shaped. Allophane has a surface morphology with irregular spherical granules with a rather dark color and spread evenly [8]. Allophane-like shaped chunk more than the natural allophane. This is because in the natural allophane still contains other minerals that may affect the structure of the surface morphology.

3.4. Allophane-like adsorption capacity to heavy metals Pb

Allophane synthesis adsorption capacity test of synthesized Allophane was conducted to determine allophane-like ability to absorb heavy metals lead (Pb). Allophane-like used are allophane-like composition of Al/Si 0.5; 0.75; 1; 1.25 and 1.5.

Factors that affect the adsorption of which is the contact time between the adsorbent and the adsorbate and the level of acidity (pH). In this study the variation of contact time and pH used to determine the optimum conditions of adsorption of heavy metal ions Pb by allophane-like. Variations contact time used is 30, 60, 90 and 120 minutes, while the pH variation is used 3-7. Fixed variables used are the number of adsorbent and adsorbate volume. Pb adsorption analysis is based on the concentration of the absorbed.

In the performance test diagram allophane-like, it can be seen that the higher the pH, the average adsorption capacity has increased to the optimum conditions and it will fall back. In each ratio Al/Si pH optimum adsorption of heavy metal ions Pb is at pH 4. This is supported which states that the adsorption of heavy metal ions Pb maximal at pH 4. At low pH (< 4) there will be protonated resulting in the H_3O^+. This will lead to competition between Pb^{2+} with H_3O^+ to bind to the negative sites adsorben so that absorption of adsorbate will be disrupted. The resulting adsorption capacity has a smaller number. At high pH, there will be $Pb(OH)_2$ so that the absorption of the adsorbent will be reduced then the number of the adsorption capacity will decrease [9].

In this study, the respective ratios Al/Si has the optimum contact time at 30 minutes. Contact time is the time required by adsorbent to absorb heavy metal ions in the adsorbate. Pb metal ions will stick to the surface of the adsorbent until the optimum contact time. When the contact time was enough or superheated then Pb metal ions cannot be taken by the adsorbent surface and be released back into solution. This will result in reduced adsorption capacity value or decreased.

Adsorbent with the most optimum absorption is at ratio of Al/Si 1.5 to pH 4 and a contact time of 30 minutes. Adsorption capacity obtained was 1.6160 mg/g, which means as much as 1.6160 mg Pb ions are absorbed from 0.01 grams allophane-like are used. The concentration of metal ions Pb absorbed an average of 1.6160 ppm of 2 ppm concentration of the initial solution in 10 mL of solution with the percentage of metal ions Pb adsorbed is 80.77%. Percentage absorption of heavy metal ions Pb using allophane-like in optimum condition has better results when compared with the adsorption of metal ions Pb with natural allophane, which has a percentage of 46.90% [10]. This is because the presence of these minerals of natural allophane so the ability of adsorbent is less.

Allophane-like with optimum adsorption results were analyzed to determine the surface area. The surface area is showed in the number of total surface area of the sample in the form of powder in each sample mass. The larger of the surface area, the greater the absorption of heavy metal ions is. In this study, allophane-like with maximum adsorption results is in ratio Al/Si 1.5.

Allophane-like has a surface area of 201.558 m^2/g. When compared with natural allophane, allophane-like has a greater surface area than the natural allophane has a surface area of 85.529 m^2/g. The difference in the surface area of natural allophane and allophane-like is 116.029 m^2/g. This is what causes adsorption of allophane-like is greater than the natural allophane.

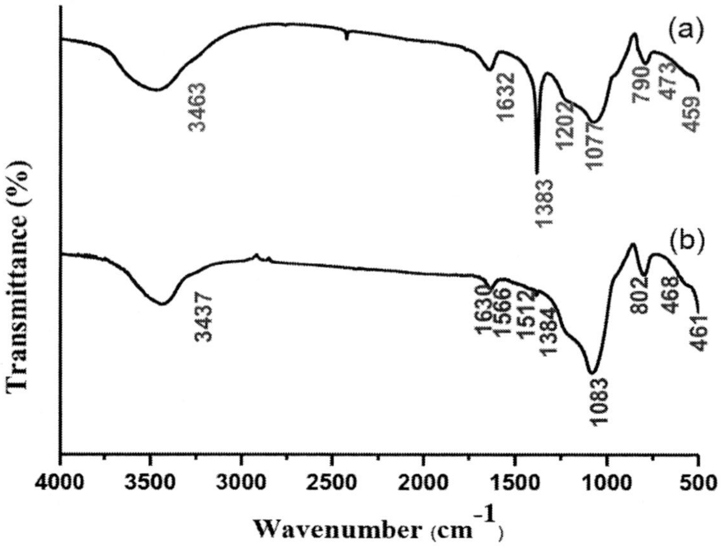

Figure 2. Analysis of the function groups before (a) and after (b) adsorption

Allophane-like before and after the adsorption shifted in 1383 into 1384; 1427; 1512; 1586 and 802 cm^{-1} which indicate a shift in the group O-Si-O or O-Al-O. The shift guessed because the hydrogen bonds between the silanol or aluminol with an adsorbent. Shifting absorption hydroxyl group (OH) was 3463 cm^{-1} to 3437 cm^{-1}. The formations of hydrogen bonds lead to the lower binding energy group O-H on Si-OH or Al-OH. The shift in the wave number of hydroxyl groups is also caused by the bonds between Pb metal with oxygen of the hydroxyl group. The metal that has been absorbed by the group Si-OH or Al-OH in the adsorbent will result in an increase of length and falling bond between the binding energy of the oxygen ions with hydrogen ions (O-H bonds) that causes the bond between them became weaker [11].

Active sites owned allophane Si-OH/ Al-OH, -O-Si-O-/-O-Al-O- and Al-O-/Si-O-. Adsorption occurs when the allophane-like oxygen ions on the surface of Pb allophane binding so that it will form a bond Si-O-Pb or Al-O-Pb. Pb which acts as a Lewis acid will receive a free electron pair of O which acts as a base lewis. The electron paired by O and Pb will form a coordinate covalent bond. In the active group -O-Si-O-/-O-Al-O- and Al-O-/ Si-O- will donate a lone pair on the Pb so that it will form a chemical bond well. The more oxygen ions on the surfaces, the more metal ions Pb absorbed by the adsorbent. This will make the adsorption capacity of greater number.

The determining of type of adsorption is done by determining the adsorption isotherm. The adsorption process according to the Langmuir showed that the interaction between Pb with allophane surface is limited to a single layer formation. Freundlich isotherm occurs in physics where the intermolecular force is greater than the force between molecules. In physics adsorption, heavy metals Pb is not strongly bound by allophane-like active sites that would cause re-release or desorption. Interactions occured by heavy metals Pb enters through the pores of allophane-like and interacting with a weak binding energy van der Waals bonding. The weak force causes adsorbat can move from one surface to the other surface of the adsorbent. To find the equation corresponding to the Langmuir or Freundlich isotherm it can be seen from the correlation coefficient (R2) of the respective linear graph. The R2 score close to 1 means that the types of adsorption isotherms tend to follow them. Based on the results of linear regression in this study, Langmuir isotherm has a score of R2 0.544 with line equation $y = 0.192x + 3.252$ while Freundlich isotherm has score of R2 0.974 with line equation $y = 0.867x - 0.529$. The R2 on Freundlich isotherm is close to 1 and higher when compared to the value of R2 in the Langmuir isotherm that the adsorption type of research is likely to follow Freundlich isotherm or physics.

4. Conclusion

The synthesis of allophane-like has successfully done with a solution of TEOS and nonahidrat (Al $(NO_3)_3.9H_2O$) based on variations in the ratio of Al/Si 0.5; 0.75, 1, 1.25 and 1.5. Allophane-like characters have functional groups -OH, O-Si-O/O-Al-O and Si-O/Al-O. Allophane-like has a low crystallinity (amorphous). Allophane-like has active sites Si-OH/ Al-OH, -O-Si-O-/-O-Al-O- and Al-O-/Si-O- which able to bind Pb metal ions. The adsorption capacity toward Pb heavy metals is shown with optimum adsorption capacity of 1.61 mg/g in the ratio Al/Si 1.5 at pH 4 with a contact time of 30 minutes. Allophane-like has isotherm types that tend to follow the type of Freundlich adsorption isotherm.

Reference

[1] Munir 1996 *Tanah-Tanah Utama Indonesia* (Jakarta: Pustaka Jaya)
[2] Peelletier E M, Bogenez S, Pelletier M, Razafitianamaharavo A, Ghanbaja J, Lartiges, and Michot L 2005 *Colloids Surf.* **255** 1-10
[3] Denaix L, Lamy, Isabelle and Bottero J Y 1999 *Colloids Surf., A* **158** 315-325
[4] Reinert, Laurence, Ohashi F, Kehal M, Bantignes J, Goze-Bac and Duclaux L 2011 *Appl. Clay Sci.* **54** 274-270
[5] Egashira K and Aomine S 2011 *Clay Sci.* **4** 231-242
[6] Abidin Z, Matsue N and Henmi T 2008 *Clays Clay Miner.* **28** 285-294
[7] Marsuki F 2014 *Skripsi Jurusan Kimia* (Surakarta: Fakultas Matematika dan Ilmu Pengetahuan Alam Universitas Sebelas Maret)
[8] Meima J A, Weijen R D, Eighmy T T, Comans R N J 2002 *Appl. Geochem.* **17** 1503-1513
[9] Baidho and Zahroh E 2013 Prosiding SNST Tahun Keempat Universitas Wahid Hasyim Semarang 43-47
[10] Pranoto, Suranto, Sugiyarto and Ashadi 2013 *J. Environ. Earth Sci.* **3** 48-56
[11] Yuanita L 2009 *Jurnal Ilmu Dasar* **10** 49-55

IOP Conf. Series: Earth and Environmental Science **75** (2017) 012016 doi:10.1088/1755-1315/75/1/012016

Design and Properties of Thin Surfacing Hot Mix Asphalt Containing Crumb Rubber as Partial Aggregate Replacement

Ary Setyawan[1], Nugroho Febrianto[2] , and Djoko Sarwono[1]
[1]Roadmate Research Group, Sebelas Maret University, Jl. Ir. Sutami 36 A Surakarta, Indonesia
[2]Civil Engineering Department, Sebelas Maret University, Jl. Ir. Sutami 36 A Surakarta, Indonesia

E-mail: arysetyawan@staff.uns.ac.id

Abstract. Road damage caused as a result of the traffic load and environment. One method to improve the road condition is from an overlay. But the new layer on the top of the pavement structure is thick enough and elevate the surface of the pavement, so it will cause some impact on the user safety and engineering. The use of a thin layer of hot mix asphalt is an alternative to anticipate the thickness problem. Crumb rubber is a waste material that has a flexible nature, these materials are used as an aggregate replacement in the hot mix asphalt thin layer. The research was conducted to find the optimum bitumen content and optimum crumb rubber content on asphalt mixtures by the Marshall procedure. Finally, it was concluded that the addition of crumb rubber in a thin layer of hot mix asphalt indicates the better the interlocking between aggregates so that gave the better Marshall stability, the higher the flow rate, the lower the marshall quotient, reduce the void ratio. The results show that the addition of crumb rubber content as an aggregate replacement leads to the use of less optimum bitumen content.

1. Introduction

The use of a thin surfacing of hot mix asphalt is one alternative that can be used to anticipate the problems of pavement surface elevated due to conventional overlay. Thin surfacing hot mix asphalt is a technology that is being developed as a preventive maintenance effort and resurfacing of the surface damaged pavement. The surface layer has a thickness that ranges between 25-40 mm and it is expected to solve problems such as fretting (loose of aggregate), more resistant to water penetration (impermeability) and increasing the roughness (skid resistance) [1-4]. The use of crumb rubber as a mixture of asphalt is also expected to reduce the use of excessive petroleum asphalt, so as to save the asphalt which is the natural resource that could not be renewable. Crumb rubber can be obtained easily because the residual material of the tire retreading industry, so indirectly to make an effort to recycle unused materials [5]. Previous researches have been conducted to assess the compatibility of using crumb rubber in the design of road materials [6-8]. Further research is required regarding the addition of crumb rubber in the design of hot asphalt mixture thin surfacing material with expectations to obtain maximum results and appropriate to the Indonesian circumstances in so it can be practically implemented.

[1] To whom any correspondence should be addressed.

2. Experimental

The materials used for this research are coarse aggregate, fine aggregate, filler, asphalt 60/70 pen and crumb rubber.The properties of coarse aggregate and fine aggregate are presented in Table 1.

Table 1. The properties of Aggregates.

No	Type of Test	Coarse aggregate		Fine aggregate	
		Results	Specification	Results	Specification
1	Adsorption (%)	2.512	max.3.0	2.093	max. 3
2	*Bulk density* (gr/cc)	2.514	min. 2.5	2.665	min. 2.5
3	SSD Density (gr/cc)	2.577	min. 2.5	2.707	min. 2.5
4	*Apparent Density* (gr/cc)	2.683	–	2.881	–

The properties of 60/70 pen asphalt were tested using Indonesian Standard Test and presented in Table 2.

Table 2. The Properties of 60/70 pen bitumen.

No	Type of Test	Value	Unit	Specification	
				Minimum	Maximum
1.	Penetration	69.7	0,1 mm	60	79
2.	Ductility	>150	cm	100	–
3.	Softening Point	48	Celcius	48	58
4.	Flash Point	257.5	Celcius	200	–
5.	Burning point	325	Celcius	200	–
6.	Specific Gravity	1,038	g/cc	1	–
7.	Affinity	98	%	95	–

The Standard manufacture Job Mix Asphalt aggregate substitute referring to Standard Specification Construction of Transport Systems (North Carolina Department of Transportation), while the testing of the specimen based on standards issued by the Asphalt Institute Superpave Series 1 (SP -1) and adopting of the method standardized by the Directorate General of Highway in the form of National Indonesian Standard (SNI).

This research was conducted by testing asphalt as a preliminary test and testing of the mixture of the characteristics of Marshall. Marshall test conducted to obtain the density, porosity, stability, flow and Marshall Quotient can then be calculated and obtained their binder content and optimum levels of addition of crumb rubber. The gradation of thin surfacing is presented in Table 3 [1].

Tabel 3. North Carolina Thin Surfacing Gradation.

No	Sieve Size (mm)	Specification	Median	Gradation Used
1	3/4" (19 mm)	100	–	–
2	1/2" (12.7 mm)	85 – 100	92.5	92.65
3	3/8" (9.51 mm)	60 – 80	70	69.30
4	No.4 (4.76 mm)	28 – 38	33	33.62
5	No.8 (2.38 mm)	19 – 32	25.5	25.16
6	No.50 (0.297 mm)	8 – 13	10	10.60
7	No.200 (0.074 mm)	4 – 7	5.5	5.68

The Crumb rubber used in this study was obtained from the retreading industry in Solo, Indonesia. The size of crumb rubber used is passed through sieve # 50 and retained by sieve # 100 and has a specific gravity of 1.058 g/cc. Figure 1 shows the crumb rubber used in this investigation.

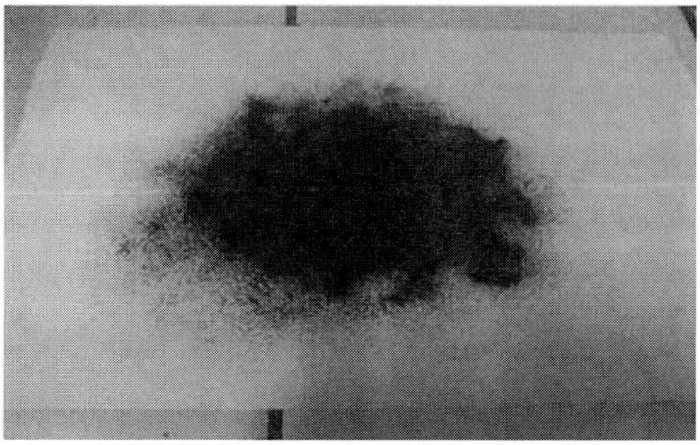

Figure 1. Crumb rubber as aggregate replacement made from waste recycled tires

3. Results and Discussion

The results of the investigation are presented in term of the influence of percentage crumb rubber as an aggregate replacement on the Marshall and volumetric properties and the influence of crumb rubber content on the optimum bitumen content of the thin surface hot mix asphalt.

3.1. The influence of crumb rubber aggregates replacement content on the Marshall and Volumetric Properties

The Optimum bitumen content was determined by performing a test Marshall. Marshall test performed based on the estimated level of asphalt while the asphalt content variation of 3.5%, 4%, 4.5%, 5% and 5.5%. It also determined the optimum level of crumb rubber to be used in research using the total volume ratio of aggregate to be used. Before the Marshall test done, first tested the Volumetric include measurement of diameter, thick and heavy in the air, then do the calculations to obtain density values, SGmix, and porosity. Marshall then new testing and obtained the value of stability, flow and Marshall Quotient. Of the value - the value can be determined mix of the best properties or the optimum bitumen content. The volumetric and Marshall properties of hot mixture thin surfacing asphalt with crumb rubber partial aggregate replacement are presented in Table 4.

The influence of bitumen contents and crumb rubber content variations are presented in Figure 2, it shows that the addition of crumb rubber in the mixtures gave influence on the value of stability. It is seen that the stability has increased with the addition of crumb rubber by 0.3% and 0.5% are 742.22 kg and 766.540 kg, respectively. This happens because the greater the crumb rubber contained in the mixture to make power tie between aggregate the better because crumb rubber will serve as a and additional binder to obtain better interlocking and bonding between aggregate and bitumen hence increasing the value of stability.

The influence of aggregate replacement by crumb rubber on the volumetric properties is illustrated in Figure 3. It could be seen that the addition of crumb rubber influence on the void ratio of the mixture. Porosity is affected by the density and the specific gravity of the mixture. It is seen that the addition of crumb rubber decreases the void ratio. This suggests that increased levels of crumb rubber by 1% to 0.5% resulted in an air cavity contained in the mixture decreases due to that the crumb rubber will serve as a binder that filled the voids as well as assist the bonding between the asphalt aggregate. This also can improve the strength of the mixture.

Table 4. The results of Marshall test at various bitumen and rubber content.

Bitumen Content	Stability (kg)	Porosity (%)	Flow (mm)	Bulk Density (g/cm^3)	*Marshall* Quotient (kg/mm)	Rubber Content (%)
3.5	410.804	11.747	1.833	2.06	223.83	
4	638.134	9.019	2.117	2.10	306.59	
4.5	750.458	8.315	2.267	2.09	330.00	0.5
5	631.445	5.043	1.600	2.14	398.17	
5.5	402.362	5.109	1.467	2.11	277.22	
3.5	489.395	12.135	1.567	2.04	322.15	
4	624.003	9.415	2.333	2.12	274.90	
4.5	766.540	8.710	2.400	2.10	320.18	0.3
5	552.027	5.447	2.200	2.09	161.53	
5.5	428.861	5.509	2.033	2.11	210.52	
3.5	421.221	12.524	1.600	2.04	258.64	
4	554.779	9.811	2.050	2.10	268.93	
4.5	716.528	9.105	1.900	2.10	374.54	0.1
5	594.695	5.852	2.100	2.10	286.66	
5.5	512.953	5.909	1.850	2.10	279.21	
5	515.301	12.718	2.250	2.09	244.93	
5.5	612.879	10.010	2.200	2.08	286.34	
6	742.223	9.302	2.183	2.11	352.46	0
6.5	479.091	6.054	2.167	2.06	220.85	
7	388.374	6.109	2.167	2.09	179.86	

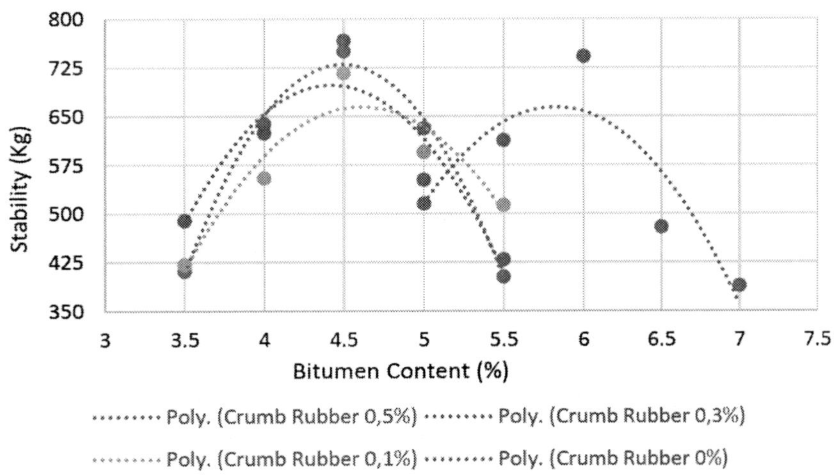

Figure 2. The Stability of thin surface asphalt mixture at various bitumen and rubber contents

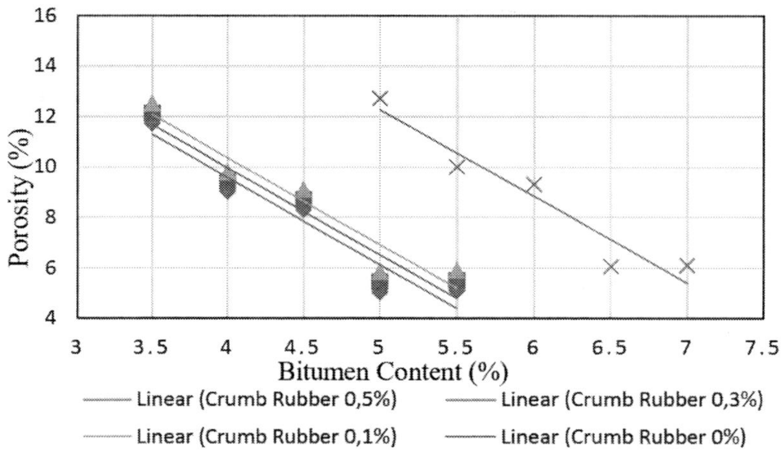

Figure 3. The porosity of thin surfacing hot mixture at various crumb rubber aggregate replacement

3.2. The influences of crumb rubber aggregate replacement on the optimum bitumen content

The optimum bitumen content of asphalt level that will produce the best characteristics in terms of the nature of the value of stability in an asphalt mixture. The bitumen content to be used in calculating the level of asphalt manufacture the next test object. To search for the value of the optimum bitumen content calculated the regression equation bitumen content relationships with stability as presented in Table 2. From the chart it can be concluded that the decreased use of crumb rubber asphalt penetration 60/70 along with increasing levels of CR is used. This happens because with greater levels of CR contained in the mixture to make asphalt easier to bind the aggregate in the mix because CR will function as a binder. The correlation between crumb rubber content and optimum bitumen content could be expressed as y = 12,96x2 - 8.692x + 5656 with $R^2 = 0.873$ as presented in Figure 4. It can be seen that the use up to 0.3% by weight of crumb rubber on the pavement mix has the effect on reducing the use of conventional petroleum bitumen hence become more environmentally friendly infrastructure construction.

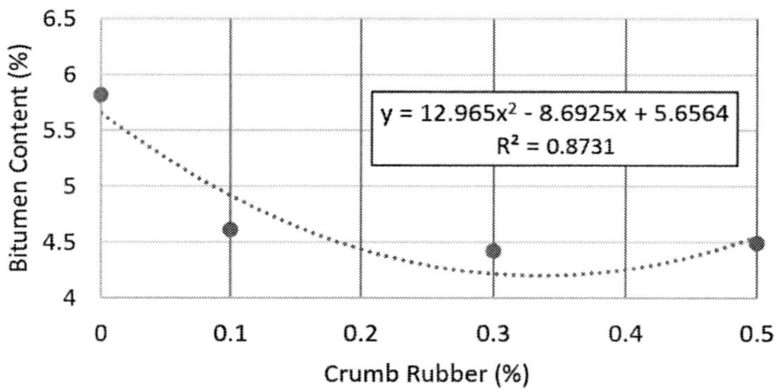

Figure 4. The influenced of partial aggregate replacement by crumb rubber on the optimum bitumen content

4. Conclusion

The addition of crumb rubber in thin surfacing hot mixture asphalt demonstrates increasing the value of Marshall Stability, Marshall Flow and decreasing the value of Marshall Quotient, porosity, hence giving better performance of the mixture. In addition, from the correlation analysis between the percentage of aggregate replacement by crumb rubber and the optimum bitumen content has the effect on reducing the optimum bitumen content, so reducing the use of conventional petroleum bitumen and become more environmentally friendly infrastructure construction.

References

[1] Uzarowski L, Maher M and Farrington G, 2005 in *Proceedings*
[2] Nicholls J C, Carswell I and Langdale P C 2002 *Transport Research Laboratory Report* 557
[3] Newcomb D E and Hansen K R 2006 in *Proceedings*
[4] Gilbert T M, Olivier P A and Galé N E 2004 *CAPSA'04* **12** 16
[5] Irman E 2014 *Proceeding Seminar Nasional Teknomedia*, 53-58
[6] Airey G D, Rahman M M and Collop A C 2003 *Journal of Pavement Engineering* **4** 105-119
[7] Yusuf M, Setyawan A, and Sarwono D 2015 *Matriks Teknik Sipil* *3* 1
[8] Ghasheir F M A B, Setyawan A and As'ad S 2016 *Jurnal Teknik Sipil* **2** 2

IOP Conf. Series: Earth and Environmental Science 75 (2017) 012017 doi:10.1088/1755-1315/75/1/012017

The Solid Phase Curing Time Effect of Asbuton with Texapon Emulsifier at the Optimum Bitumen Content

D Sarwono[1], R Surya D [2], A Setyawan[1] and Djumari[1]
[1] Roadmate Research Group, Sebelas Maret University, Jl. Ir. Sutami 36 A Surakarta, Indonesia
[2] Civil Engineering Department, Sebelas Maret University, Jl. Ir. Sutami 36 A Surakarta

E-mail: sarwono60@yahoo.com

Abstract. Buton asphalt (asbuton) could not be utilized optimally in Indonesia. Asbuton utilization rate was still low because the processed product of asbuton still have impracticable form in the term of use and also requiring high processing costs. This research aimed to obtain asphalt products from asbuton practical for be used through the extraction process and not requiring expensive processing cost. This research was done with experimental method in laboratory. The composition of emulsify asbuton were 5/20 grain, premium, texapon, HCl, and aquades. Solid phase was the mixture asbuton 5/20 grain and premium with 3 minutes mixing time. Liquid phase consisted texapon, HCl and aquades. The aging process was done after solid phase mixing process in order to reaction and tie of solid phase mixed become more optimal for high solubility level of asphalt production. Aging variable time were 30, 60, 90, 120, and 150 minutes. Solid and liquid phase was mixed for emulsify asbuton production, then extracted for 25 minutes. Solubility level of asphalt, water level, and asphalt characteristic was tested at extraction result of emulsify asbuton with most optimum ashphal level. The result of analysis tested data asphalt solubility level at extract asbuton resulted 94.77% on 120 minutes aging variable time. Water level test resulted water content reduction on emulsify asbuton more long time on occurring of aging solid phase. Examination of asphalt characteristic at extraction result of emulsify asbuton with optimum asphalt solubility level, obtain specimen that have rigid and strong texture in order that examination result have not sufficient ductility and penetration value.

1. Introduction

Asbuton Emulsion is a method developed to extract the bitumen content in Asbuton, extraction results obtained is a material which has some similarities like the nature of the bitumen in the bitumen crude oil processing residue results. The study ever conducted on asbuton extraction merely reviewing the content / composition of the constituent material optimum asbuton bitumen emulsion to produce the highest levels [1, 2]. Therefore, this study seeks to get the best asphalt content by optimizing the reaction or bonding that occurs between asbuton and premium with aging or keep a mixture of such materials in a container that has been prepared. Fluxing oil is liquid element whose function to soften the asphalt on asbuton. Texapon emulsifier is an emulsifier solution which serves to make an emulsion of oil, water and Asbuton so that such substances can be mixed in the form of slurry. Grinder 60- MB type rotation

[1] To whom any correspondence should be addressed.

speed of 3000 rpm, is used as a means of mixing the solid phase mixing between asbuton grain 5/20 with premium fluxing.

2. Experimental

This study was carried out experiments to obtain the extraction asbuton emulsion to be further advanced research on materials extraction results. Making the material extracted utilizing the process of emulsion (Asbuton emulsion) using emulsion method. Emulsion method for the separation of asphalt on asbuton of mineral require emulsifier that is similar to the processing of bitumen emulsion. The process of making the test object, there are two phases, namely solid and liquid. Solid phase is a mixture of grains asbuton 5/20 with premium fluxing mixed using a grinder-type MB 60 with mixing time 3 minutes. The solid phase mixture is then processed at the aging stage solid phase with a variation of curing time 30 minutes, 60 minutes, 90 minutes, 120 minutes and 150 minutes. Varian time on curing the solid phase is expected to affect the values obtained from the bitumen content of asphalt solubility assay. While the liquid phase consists of texapon as an emulsifier, HCl and distilled water. Texapon acts as an emulsifier for a process of oil in water emulsions. Solid phase and liquid phase mixed using a drill mixer modification spin speed 2000 rpm. Then the extraction process on a machine centrifuge extractor for 25 minutes 20x100 rpm speed. Results will separate bitumen extraction of minerals in Asbuton. Minerals will be left behind in the bowl while the extraction of bitumen emulsion machine will be out of the bowl and galvalum accommodated in containers that have been prepared. Asphalt emulsions are accommodated in galvalum container extraction h then conducted the test include: the levels of solubility of hydro-carbon, water content and characteristics of the asphalt.

The procedures of sample producing are as follow:

1. Grinding the grain 5/20 asbuton the clot due to long storage time using a grinder-type players MB 60 with blunt knives.
2. Preparation of solid phase extraction asbuton which is a mixture of grains asbuton 5/20 and fluxing (premium) by weight according to plan. Weight premium used was 133.33 grams, equivalent to a ratio of 1: 3 by weight asbuton. Mixing grains asbuton 5/20 of 400 grams and premium gasoline using a drill that has been modified with a stirrer while mixing solid phase for 3 minutes. Drill rotation speed is 2000 rpm.
3. When the solid phase is completed, then perform the curing of the solid phase results in the porcelain dish that has been prepared with a variety of curing time 30,60,90,120 and 150 minutes.
4. Preparation of a liquid phase which comprises the material texapon (emulsifier), HCl and distilled water. Weight texapon that by 3.13 grams, 4.80 grams HCl and distilled is 200 grams. Then the liquid phase is allowed to stand for ± 24 hours for the foam formed during mixing of the liquid phase disappears and liquid return.
5. Some modifications to the extraction machine centrifuge extractor is done by adding galvalum container as the container extraction asbuton fine emulsion, sieve # 60 which is placed between the bowl and cover the bowl, as well as a double perforated clamp plate galvalum sieve # 60.
6. The extraction process asbuton emulsion for 25 minutes, at 20x100 rpm rotation speed. Bitumen extraction process separates the content (in the form of emulsions) and minerals contained in asbuton emulsion.

The test object is made triple means three test object will have the weight of the composition and method of making the building blocks of the same on each of these variants curing time. The goal is that the average yield of the three objects of the same test to ensure the accuracy of test data. Obstacles encountered is when mixing the solid phase, Grinder type MB 60 is not able to maintain the current composition of the mixture statutes mixing process runs. As a result, a mixture of solid phase bounced out and attach to almost the entire wall where mixing the Grinder. As a solution Grinder MB type 60 still function but with different uses of the initial planning. Table 1 presents the planning of each materials composition.

Table 1. The Materials composition

Code of sample	Granular Asbuton (gram)	Premium (gram)	Time of aging (menit)	*Time of Mixing* (menit)	HCl (gram)	Texapon (gram)	Aquades (gram)	Time of Ekstration (menit)
1A			30					
1B			60					
1C	400	133.33	90	3	4.80	3.13	200	25
1D			120					
1E			150					
2A			30					
2B			60					
2C	400	133.33	90	3	4.80	3.13	200	25
2D			120					
2E			150					
3A			30					
3B			60					
3C	400	133.33	90	3	4.80	3.13	200	25
3D			120					
3E			150					

3. Results and Discussion

3.1. Testing of asphalt solubility

Assay solubility of asbuton emulsion bitumen extraction results in this study was calculated using a filter paper and the solvent is TCE (trichlorethylene). The method refers to the ISO 2438: 2015 on the solubility test method for asphalt [3,4].

Table 2. The solubility of Asphalt

Code of sample	Time of aging (minute)				
	30 (%)	60 (%)	90 (%)	120 (%)	150 (%)
P1	91.50	95.48	94.23	96.76	95.63
P2	90.10	93.75	93.60	94.69	94.03
P3	91.04	91.18	92.49	92.86	90.65
Average	90.88	93.47	93.44	94.77	93.44

From Table 2 can be graphed that solid phase curing time relationship with the levels of solubility in asbuton bitumen emulsion. The graph of the relationship can be seen in Figure 1. Based on these data, obtained coefficient determination $r^2 = 0.8863$ while the value of the correlation coefficient $r = 0.9414$, or can be said to be a very strong correlation. The longer the curing time, the solid phase will be increasing the dissolved asphalten component of Asbuton grains in the emulsion Asbuton up at a certain time has decreased. This is because at a certain time span, fluxing premium gasoline have maximum working attractive asphalt asbuton attached to rocks. In this study, certain span of time is 120 minutes. Testing of bitumen content of asphalt solubility test produces a value above 90%, which means that the content of the asphalt on the test object is high. But 90% is not entirely is pure asphalt but still leaves the other emulsion constituent ingredients such as premium, HCl, texapon and distilled water.

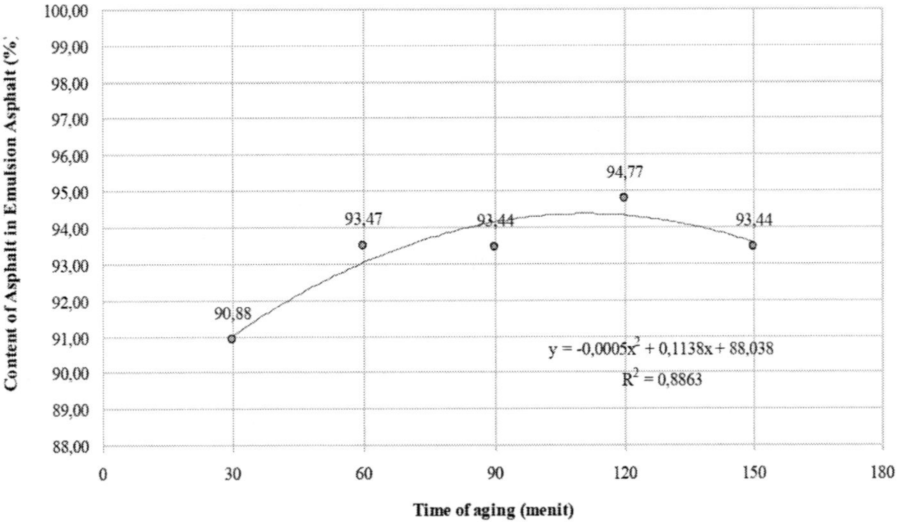

Figure 1. Aging time relationship graph with levels in Buton Bitumen Emulsion

Bitumen content extraction results obtained researchers increased dramatically compared to the secondary data from the factory in the range of 20% and research on the extraction of emulsions asbuton previous range 26-32% [5]. This happens because the development of methods for processing and modification of the research methods of extraction asbuton previous emulsion. Researchers used a method which enables bitumen emulsion inherent in mineral asbuton can be detached, then the process of extraction can separate the bitumen from the mineral asbuton although not in its entirety. Some modification conducted in this research is to add a sieve # 60 and galvalum perforated plate so that the mineral held by sieve # 60 does not come out of the bowl when the extraction process takes place. This is a significant impact on support levels to drop filler content in the extraction asbuton emulsion. This is evidenced by the results of solubility test of asphalt that produce high levels of above 90%.

3.2. Water Content

Test the water content of the emulsion asbuton extraction results are basically the same with the test of water content [5].

Table 3. Water Extraction Results from Asbuton Emulsion

Time of aging (minute)	Water content P1 (%)	Water content P2 (%)	Water content P3 (%)	Water content average (%)
30	5.00	6.53	6.78	6.10
60	4.27	4.11	5.02	4.47
90	4.01	5.68	4.49	4.72
120	4.23	4.50	4.83	4.52
150	4.20	4.36	5.54	4.70

Results graph curing time of the solid phase of water content in asbuton emulsion (AE) is presented in Figure 2

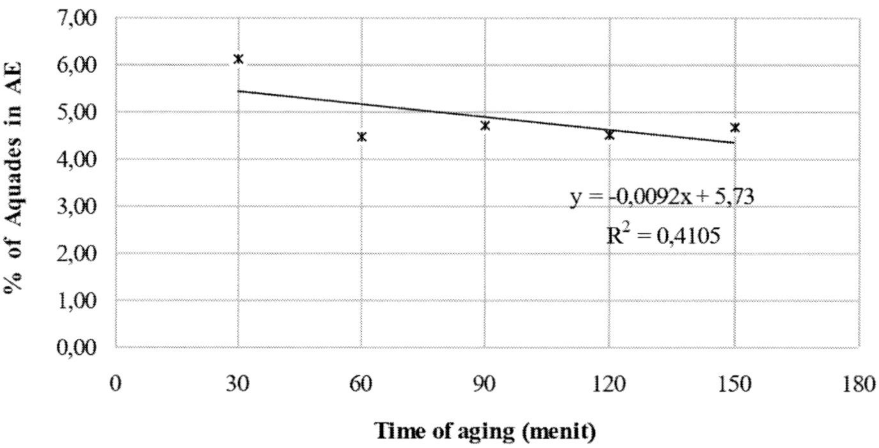

Figure 2. Relationships graph between Curing Time and Water Content of the Emulsion Extraction Asbuton

Figure 2 gives an overview obtained coefficient determination $r^2 = 0.4105$ while the value of the correlation coefficient of 0.6407, showing a correlation between curing time with the content of water in emulsion Asbuton extract strong enough. From the regression analysis can be expressed the longer the curing time of solid phase it will obtain the water content in the extraction asbuton emulsion get smaller, it is because the longer the curing time of solid phase (asbuton grain and premium) will make the asphalt bound by fluxing more. The function of the liquid phase (texapon, HCl and distilled water) to separate a mixture of asphalt and fluxing premium of mineral increasingly needed in large numbers. Then the water content is still contained in the emulsion decreases in line with the longer curing time the solid phase.

3.3 Test Results of Asphalt Emulsion Extraction Characteristics

Testing of bitumen characteristics is intended to determine the value of the test ductility, penetration, softening point, flash point and fuel, density and viscosity.

Table 4. Testing Results of Asphalt Emulsion Asbuton Extraction Characteristics

Type of Test	Test Standard	Results	Unit
Ductility	SNI 06-2432-1991	- *	cm
Penetration	RSNI 06-2456-1991	- *	10^{-1}mm
Softening Point	RSNI 06-2434-1991	>100	°C
Flash and burn point	SNI 06-2433-1991	270 and 290	°C
Specific Gravity	SNI 06-2441-1991	1.66	g/cc
Affinity	PA-0312-76	100	%

Table 4 is a picture of the nature of the asphalt properties (after the emulsion process is completed) on the extraction asbuton asphalt emulsion that has the most optimum levels. Based on these test results, obtained the value of ductility and penetration test has not been measured, as well as the test value is very high softening point. The main factor that makes the absence of penetration and ductility values are used, namely fluxing premium [6]. This occurs because the premiums as a fluxing participate in asbuton evaporated when the emulsion is completed as the evaporation of moisture present in the emulsion. Selection of premium gasoline as a fluxing caused by the extraction asbuton previous research results that fluxing used is kerosene, takes quite a long time (months) to experience settlement / precipitation emulsion mixture. The purpose of the use of premium gasoline as a fluxing in order to speed up the settling time of bitumen. But unanticipated nature of the gasoline that is easy to evaporate so that the components oil / oil on asphalt go missing in line with the completion of the process Asbuton emulsion back into Asbuton and obtained a very hard asphalt texture. The process of evaporation of water is common in the asphalt emulsion, (heating step to remove the element of water in the emulsion be taken to accelerate the process in order to test the characteristics to do). Texture hard and stiff obtained after missing water content indicates that not only the element of the water is evaporated oil, but the elements also disappear given the premium nature volatile. The results of this product or test specimen is not suitable when used as a binder in the mix of the surface layer (surface course) of pavement due to a very high stiffness. But the results of this product can be used for a mix function as a foundation layer (base course), and developed into the added material or modifications to conventional asphalt (bitumen 60/70 pen) for increasing performance and to reduce the amount of consumption of crude oil and bitumen to optimize asphalt Asbuton are not so utilized [7,8].

4. Conclusion

Asphalt optimum levels generated from bitumen solubility is equal to 94.77% in the variation of the solid phase curing time of 120 minutes. The levels of distilled water in the extraction asbuton emulsion showed a decrease in the water content of the emulsion asbuton the longer curing time solid phase takes place. The test results from the extraction of bitumen characteristics asbuton emulsions, suggesting that the fluxing premium (petrol) participated evaporate / lost with the completion of the process Asbuton emulsion becomes Asbuton extract

References

[1] Affandi F 2008 Pusat Litbang Jalan dan Jembatan.
[2] Anonymous 1989 Revisi SNI 03-1737-1989 (Bandung: Badan, Litbang Departemen Pekerjaan Umum)
[3] Bina Marga 1999 *Pedoman Pembuatan Aspal Emulsi Jenis Kationik*. Pedoman Teknik No: 024/ T / BM / 1999 (Jakarta: PT. Mediatama Saptakarya)
[4] Ditrektorat Jendral Bina Marga. 2006. Interim Guidance No: 001 – 01 / BM / 2006. Jakarta, Kementrian Pekerjaan Umum
[5] Buana R P M, Sarwono D and Djumari D 2013 *Matriks Teknik Sipil* **1** 557
[6] Ardhyanto D, Sarwono D and Setyawan A 2013 *Matriks Teknik Sipil* **1**
[7] Annas P L, Sarwono D and Djumari D 2013 *Matriks Teknik Sipil* 1 440
[8] Eka N S J, Sarwono D and Setyawan A 2014 *Matriks Teknik Sipil* **2** 9

Improved of Natural Gas Storage with Adsorbed Natural Gas (ANG) Technology Using Activated Carbon from Plastic Waste Polyethylene Terepthalate

Yuliusman[1], Nasruddin[2], A Sanal[1], A Bernama[1], F Haris[1] and M Hardhi[1]

[1]Chemical Engineering Department, University of Indonesia, Kampus Baru UI Depok, Indonesia
[2]Mechanical Engineering Department, University of Indonesia, Kampus Baru UI Depok, Indonesia

Email: usman@che.ui.ac.id

Abstract. Indonesia imports high amount of Fuel Oil. Although Indonesia has abundant amount of natural gas reserve, the obstacle lies within the process of natural gas storage itself. In order to create a safe repository, the ANG (Adsorbed Natural Gas) technology is planned. ANG technology in itself has been researched much to manufacture PET-based activated carbon for natural gas storage, but ANG still has several drawbacks. This study begins with making preparations for the equipment and materials that will be used, by characterizing the natural gas, measuring the empty volume, and degassing. The next step will be to examine the adsorption process. The maximum storage capacity obtained in this study for a temperature of 27°C and pressure of 35 bar is 0.0586 kg/kg, while for the desorption process, a maximum value for desorption efficiency was obtained on 35°C temperature with a value of 73.39%.

1. Introduction

Fuel Oil is the main form of fuel used by Indonesian people for transportation. Usage of Fuel Oil consumption in Indonesia increases because of the rapid growth of motor vehicle industry, around 5-6% annually [1]. Indonesia imports Fuel Oil in a huge amount, reaching 1.06 million tons or worth US$ 1.13 billion [1]. Based on such issue, natural gas has a large potential to replace Fuel Oil as Indonesia possesses plentiful amount of natural gas reserve. Indonesia is the 14th highest country in the world for amount of natural gas reserve, with the reserve amounting to 103.3 trillion cubic feet [1].

The main issue in the usage of natural gas as fuel lies within the storage and distribution process, as under normal conditions natural gas is in the gas phase, storage capacity would become small and inefficient to be used. Currently, the method mainly used for natural gas storage is Compressed Natural Gas (CNG) and Liquefied Natural Gas (LNG). However, CNG technology requires very high pressure, which is > 250 bar. On the other hand, LNG technology requires extremely low temperature (- 161°C). Both of these technology have shortcomings in the technical and economic aspects. To overcome these issues, an innovation of natural gas storage is needed. Adsorbed Natural Gas (ANG) technology is a technique to store natural gas in a cylinder, through usage of activated carbon as adsorbent. This technology allows natural gas storage on a low pressure around 34 bar while still having high storage capacity [9].

Activated carbon as and adsorbent can be manufactured from the plastic waste polyethylene terephthalate (PET) for its abundance in Indonesia. The amount of polyethylene terephthalate waste generated each day reaches 26,500 tons [11]. Aside from statistics, activated carbon manufactured from PET possesses the highest surface area compared to PVC (polyvinyl chloride) and waste tires [9].

Unfortunately, ANG technology has not been applied much in motor vehicles although it has more advantages compared to CNG. One of the reasons for such cause is the adsorbent's capacity to adsorb natural gas. To improve the performance of ANG technology, PET-based activated carbon is used as adsorbent to find the optimum condition during adsorption process. Performance of ANG is attributed to several factors, which are the temperature and pressure of the operation. Keeping in mind such factors, in this study natural gas storage will be tested with ANG technology with variations being natural gas pressure and storage temperature to generate data of natural gas storage capacity inside ANG cylinders. Activated carbon will be made of polyethylene terephthalate from previous study.

2. Experimental

2.1. Equipment scheme and research flowchart

To examine the adsorption of methane gas in adsorbed natural gas (ANG) cylinder, the first step required is degassing. An amount of activated carbon which is to be tested is put inside the ANG cylinder. Circulating thermal bath (CTB) is then operated to increase and maintain vessel temperature at 26°C. Pressure output regulator on the methane gas cylinder is set at 3000 KPa (30 bar). Methane gas is then supplied through computer control and the accumulation capacity recorded through coriolis. In this study, storage and flow capacity is recorded through the accumulation of gas in and out of coriolis with a unit of gram. Figure 1 shows the natural gas storage scheme with ANG technology.

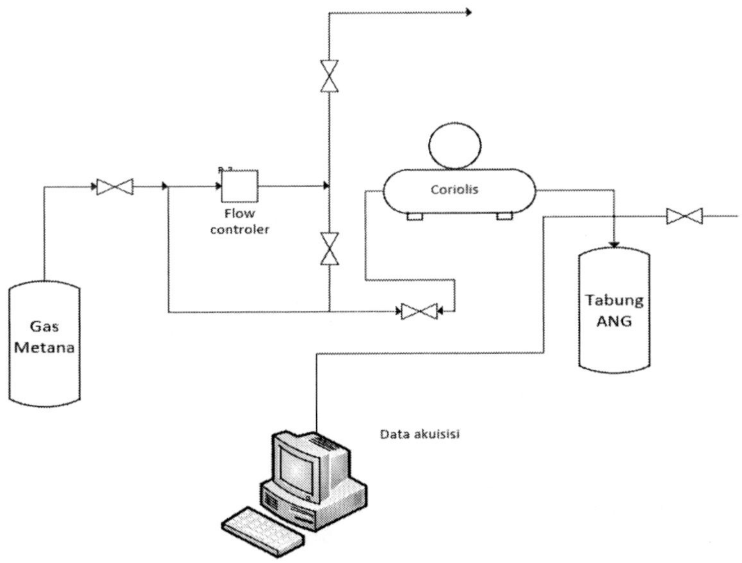

Figure 1. Schematic for storage process

2.2. Measurement of empty volume

Measurement of empty volume is conducted in order to find out the volume on measuring cell. The empty volume of the measuring cell is the total volume of empty space that existed on measuring cell.

$$V_{empty} = V_{MC} - V_{space\ filled\ with\ adsorbent} + V_{adsorbent\ pores} \tag{1}$$

Annotations:
V_{empty} = volume of empty measuring cell (m^3)
V_{MC} = volume of measuring cell filled with adsorbent (m^3)

This process begins with conditioning the charging cell at 30°C using circulating thermal bath. In the next step, helium gas is supplied to the charging cell by opening the valve until the desired pressure is achieved, while charging cell's initial condition is recorded. Helium gas is then streamed to measuring cell from charging cell until equilibrium is achieved, the time needed around 15 minutes. The charging and measuring cell's final condition is recorded as well. Data obtained during the procedure will then processed to obtain the value for empty volume.

2.3. *Adsorption test*

Study for isothermal adsorption is conducted with volumetric method. The basis for volumetric measurement is pressure, volume, and temperature in which data is recorded when adsorbate enters the chamber where adsorbent is present. After adsorption equilibrium is reached, the amount of adsorbate adsorbed is measured from the change in pressure using ideal gas equation. Mass balance for adsorbate vapor inside the charging and measuring cell can be assumed as follows [11]:

$$m_{d,mc} = |m_{cc}| - m_{ads} \quad (2)$$

Annotations:
$m_{d,mc}$ = mass of adsorbate in *measuring cell* (kg)
$|m_{cc}|$ = mass of adsorbate in *charging cell* which moved to *measuring cell* (kg)
m_{ads} = mass of adsorbate adsorbed by adsorbent (kg)

Where ρcc and ρmc are the density for adsorbate at a temperature and pressure inside charging cell and measuring cell. The value of ρcc and ρmc can be obtained by utilizing the software REFPROP.

3. Results and Discussions

3.1. *Activated carbon*

In this study, PET-based activated carbon is obtained from previous research [13]. The result for PET-based activated carbon characterization can be seen on Table 1.

Table 1. Characterisation of activated carbon

Characteristic	Value
Surface Area, Iod Value (m^2/g)	1591.72
Pores Volume (cc/g)	0.176
Micropores Volume (cc/g)	0.136
Pore Diameter (A°)	15.2
Mesh Size	10

3.2. *Natural gas composition*

Adsorbate used to examine ANG is natural gas bought from Stasiun Pengisian Bahan Bakar Gas (SPBG) Pertamina. Results for natural gas composition characterization is obtained through gas chromatography seen on Table 2.

Table 2. Natural gas composition

No	Name	%
1	CH_4	87.762
2	N_2	4.698
3	C_2H_6	3.781
4	CO_2	2.260
5	C_3H_8	1.278
6	C_4H_{10}	0.222
	Total	100.0

3.3. Measurement of empty volume

In this study, measurement of empty volume on measuring cell is conducted, which is empty space on measuring cell cylinder which has been filled with activated carbon. 5 g of activated carbon is used. Examination of empty volume is needed as it is one of the parameters to measure adsorption capacity. Measurement of empty volume done three times, the results are averaged to obtain an accurate value for empty volume. From the data, it is seen that the average value for empty volume is 274.5 cm^3, which means in the measuring cell cylinder there is 274.5 cm^3 empty space, including the pores of activated carbon inside the cylinder. The results of the measurement can be seen from Table 3.

Table 3. Measurement of empty volume

Trial		Pressure (kPa)	Temperature (°C)	Z	n	ni	V empty (cm^3)
	Initial ρcc	724.5	29.5	1.0033	346.9	65.2	
1	Final ρcc	589.1	30.0	1.0027	281.7		274.9
	ρ mc	584.1	30.1	1.0027			
	Initial ρcc	750.6	30.0	1.0026	359.1	67.2	
2	Final ρcc	610.4	30.0	1.0027	291.9		273.9
	ρ mc	605.4	30.0	1.0026			
	Initial ρcc	712.1	29.0	1.0033	341.6	64.2	
3	Final ρcc	579.9	30.1	1.0027	277.3		275.2
	ρ mc	575.0	30.1	1.0026			
			Average				274.5

3.4. Adsorption test

Adsorption test is done with variations to pressure and temperature. 5 g of activated carbon is also used on this test. The data obtained is storage capacity, obtained in natural gas weight (kg) adsorbed per activated carbon weight (kg) used. The following graph shows amount of gas adsorbed vs. pressure for PET-based activated carbon.

From Figure 2, it can be inferred as pressure increases, adsorption capacity for the natural gas also increases. This indicates that the adsorption that happens on activated carbon is a physical adsorption. On the graph, the maximum point for adsorption value happens at 35 bar pressure for a value of 0.0586 kg/kg.

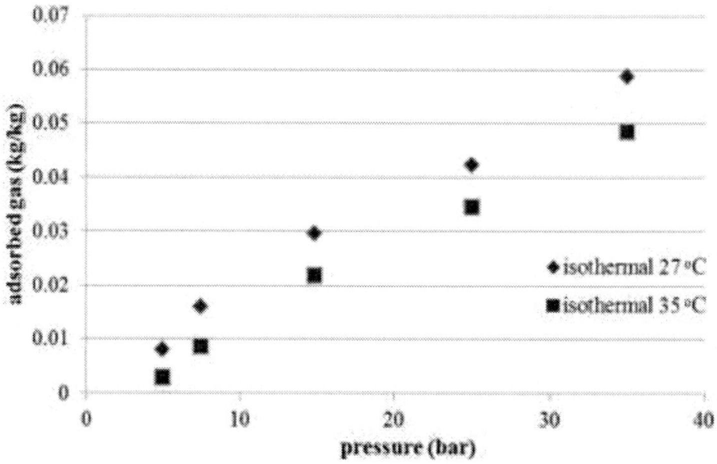

Figure 2. Change in adsorption capacity against pressure and temperature

3.5. Desorption capacity

Desorption capacity of an adsorbent is the ability of an adsorbent to release adsorbed adsorbate [4]. Desorption capacity is important as it indicates the amount of gas that could be released from the ANG cylinder. Percent efficiency here mirror the adsorbent's capacity to desorb adsorbed natural gas. Figure 3 shows, as temperature increases so is the efficiency of desorption. This happens because desorption is an endothermic process, desorption will increase if temperature is increased.

Figure 3 also shows that not all stored natural gas can be desorbed. In adsorbed natural gas (ANG) system, storage capacity is always higher than delivering capacity (desorption). Usually, storage capacity is 15 – 30% higher than delivering capacity [6]. The remaining amount of gas left inside the cylinder will depend on the type of activated carbon used, the bigger the micropores on the adsorbent, the higher the amount of remaining gas left [10]. In this study, desorption efficiency is found to increase with temperature, at 35°C giving an efficiency of 73.39%. The remaining efficiency data obtained is also in an excellent range between 70% - 85% [6].

Not all gas can be delivered, some are stuck. The mass of the stuck methane gas is caused because the adsorbate adsorbed to the adsorbent on ANG has a tendency to stay adsorbed to the adsorbent. To remove all adsorbate adsorbed off the adsorbent, an additional amount of energy is needed through heating in a high temperature.

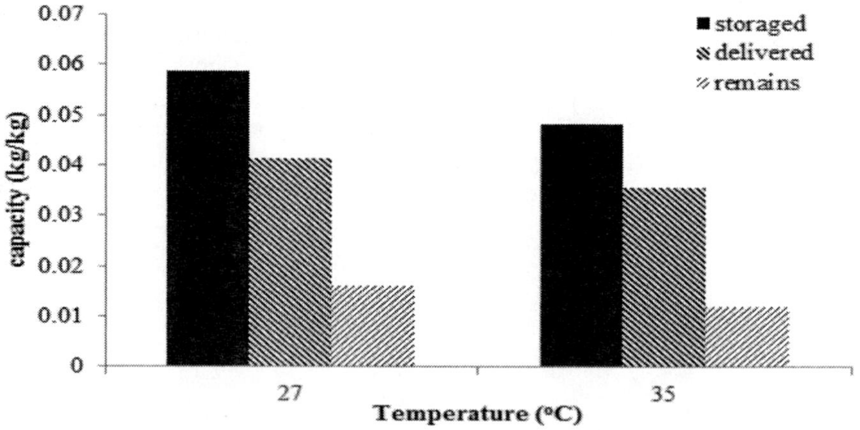

Figure 3. Effect of temperature to natural gas desorption capacity

4. Conclusions

From this study of Enhancement of Natural Gas Storage Through Usage of Activates Carbon Made from Plastic Waste Polyethylene Terepthalate, several conclusions can be made. As temperature increases, adsorption capacity will lower as adsorption is an exothermic process. As pressure increases, adsorption capacity will also rises as the amount of gas in contact with activated carbon's surface area increases. Highest adsorption capacity is reached at 27°C temperature and 35 bar pressure, with a value of 0.0586 kg/kg. As temperature increases, desorption efficiency will increases as well where at 35°C the efficiency has a value of 73.39%.

References

[1] Statistic Center of Indonesia 2014 *Tabel jumlah kendaraan bermotor.* http://bps.go.id/tab_sub/view.php? Kat =2&tabel = 1&daftar =1&id_ subyek=17&notab=12
[2] Baghery, Narges, Abedi and Jalal 2011 *Chem. Eng. Research Desi* **89** 2038-2043
[3] Bratek W, Świątkowski A, Pakuła M, Biniak S, Bystrzejewski M and Szmigielski R 2013 *J. Analytical Appl. Pyrolysis* **100** 192-198.
[4] Duong D Do 1998 *Adsorption Analysis: Equilibria and Kinetics* (London : Imperial College Press)
[5] Esfandiari A, Tahereh K and Mansooreh S 2012 *J. Taiwan Inst.Chem. Eng.* **43** 631–637
[6] Judd R W, Gladding D T M, Hodrien R C, Bates D R, Ingram J P and Allen M 2010 *BG Technology, Gas Research and Technology Centre*
[7] Lian Fei, Baoshan Xing and Lingyan Zhu 2011 *J. Colloid Interface Sci.* **360** 725–730
[8] Mat, B H, Zakaria, Zainal, Paou and Terry G 2006 *The Development of Adsorbent Based Natural Gas Storage For Vehicle Application* Universiti Teknologi Malaysia
[9] Samid, D D 2011 *Activated Carbon Characteristics Optimization System On Adsorbed Natural Gas In Dynamic Conditions,* Fakultas Teknik Universitas Indonesia
[10] Wibowo W 2015 *Pembuatan Karbon Aktif Teraktivasi KOH dari Tempurung kelapa untuk Penyimpanan Gas Metana dengan Menggunakan Teknologi Adsorbed Natural Gas (ANG),* Fakultas Teknik Universitas Indonesia
[11] Dawoud B and Yuri A 2003 *Int. J. Heat and Mass Tran.* **46** 273-281
[12] Indonesian Ministry of Environtment 2012 *Jumlah sampah plastik.* http://menlh.go.id/
[13] Yuliusman and Alristo S 2015 *Preparation of PET (Polyethylene Terephthalate) Plastic Waste Based Activated Carbon for Methane Gas Storage With ANG (Adsorbed Natural Gas) Technology,* Fakultas Teknik Universitas Indonesia

IOP Conf. Series: Earth and Environmental Science 75 (2017) 012019 doi:10.1088/1755-1315/75/1/012019

Selecting the Best Materials Compositions of Resin Based Bioasphalt

Ary Setyawan[1], **Budi Widiharjo**[2], **and Djumari**[1]
[1]Roadmate Research Group, Sebelas Maret University, Jl. Ir. Sutami 36 A Surakarta, Indonesia
[2]Civil Engineering Department, Sebelas Maret University, Jl. Ir. Sutami 36 A Surakarta

E-mail: arysetyawan@staff.uns.ac.id

Abstract. Damar asphalt is one type of bioaspal which is a mixture with the main ingredient is a resin as a binder and cooking oil as a solvent. One major drawback of this damar asphalt is the low ductility. To improve the ductility values, then use the added material Filler. Filler serves as a divider between the impurities with damar asphalt, increases ductility and increase the ability of cohesion or bonding between the particles of material damar asphalt. The purpose of this study was to determine damar asphalt modifications to the properties in accordance with the properties of damar asphalt test specifications based on the value of penetration. This method uses some variant on material such as powder bricks and fly ash as a binder. Solvent in constituent used oil and used cooking oil. It also added the polymer latex up to 10% at intervals of 2%.. The best composition of damar asphalt materials were obtained with gum rosin, Fly Ash, Oil and Latex. Damar asphalt modification damar asphalt optimum mix of resin (100g pure resin or resin chunk + 350g powder), Fly Ash powder (150g), cooking oil (205g), and latex 4%, ductility increased from 63.5 cm to 119.5 cm , the value of the flash point was originally at temperature of 240 °C to 260 °C, damar asphalt penetration of 68.2 dmm to 43 dmm, and the value of density decreases from 1.01 g / cm^3 to 0.99 g / cm^3. Damar asphalt at these modifications meet the specifications in terms of solubility in trichlore ethylene is equal to 99.5%, and also meet the affinity of damar asphalt at 99%. With the optimum value, damar asphalt could be categorized as bitumen 40/60 penetration.

1. Introduction

Increased environmental regulations and the rising costs of asphalt binder have encouraged researchers to investigate alternative binders that can be used for hot mix asphalt (HMA) [1]. The use of wasted materials has been done by incorporating wet process of Recycled Asphalt Single (RAS) and bioasphalat gave comparable results to conventional asphalt [2]. Damar asphalt have similar characters with bitumen pen 60/70 but do not meet the requirements of ductility and exceed the requirements for melting point [4]. Damar asphalt has a higher value of workability index based on the standards provided by Cabrera and Dixon is 6 [3]. Damar asphalt meet the requirements in terms of stability, density (density), flow and value marshal quotient. Resin is a natural resin produced by plants Dipterocarpaceae (genus Shorea, Hopea, Balanocarpus and Vateria) and Burseraceae (clan canarium) [15]. Asphalt composition comprised of asphaltenes which is material in black or dark brown and maltenes which is a viscous

[1] To whom any correspondence should be addressed.

liquid consisting of resins and oils. Damar is a major component in damar asphalt as a natural resin, has hardness depending on the temperature of the so-called thermoplastic properties [16]. To follow up on it, then further research on asphalt resin by changing the composition of the mixture with some variants of such material, resin, fly ash, brick powder, used motor oil, cooking oil and latex. The use of varied materials aims to get maximum results.

2. Experimental

The function is as a resin binder of damar asphalt itself. Many say the least influential resin mixture. The resin used is a resin that has been in the packaging in the form of powder [3]. Fly ash that serves as a filler and as absorbent damar asphalt work by removing impurities contained in damar asphalt mixture during cooking so that the results obtained pure mixture [5,6]. Fly is not completely lost during the cooking process, but still contained within the composition damar asphalt evident when the testing solution trichlore ethylene results of the tests showed that damar asphalt 95%, 5% insoluble fly ash is a non hydrocarbon which is insoluble in a solution of ethylene trichlore.

Cooking oil as diluent mixture damar asphalt which also play a role in reducing the elasticity damar asphalt plastic traits possessed by the resin. Latex is very influential in the rate of elastic damar asphalt evident from the results of testing ductility by adding 4% latex ductility damar asphalt meet the requirements of penetration bitumen. From the research damar asphalt the resin composition, fly ash, oil and latex fulfill some requirements as bitumen or asphalt pen 40 pen 60. Two tests that do not meet the requirements of the test is the specific weight and soluble in a solution of ethylene trichlore.

The method used is an experimental method to the test object damar asphalt with a mixture of damar resin, fly ash, and wasted cooking oil, which is used as a binder layer of asphalt concrete compared to asphalt penetration [7]. Testing standards used are:

1. SNI 06-2456-1991 (Penetration Test) [8].
2. SNI 06-2434-1991 (Softening Point Test) [9].
3. SNI 06-2433-1991(Flash Point Test) [10].
4. SNI 06-2441-1991 (Specific Gravity Test) [11].
5. SNI 06-2432-1991 (Ductility Test) [12].
6. PA-0312-76 (KVBB-V-19) (Affinity Test) [13].
7. SNI 06-2438-1991 (Solubility Test) [14].

Table 1. Material composition

Sample	Composition		
	Damar	Filler	Solvent
	450 g	150 g	205g - 230 g
S1	Damar	Red Brick	Wasted Cooked Oil
S2	Damar	Fly Ash	Wasted Cooked Oil
S3	Damar	Fly Ash	Recycled Oil

3. Results and Discussion

3.1.1. Selecting the materials composition to characterize the basic damar asphlat mixtures

Previous research has been done with various material compositions and the method of mixing [3,4]. The various compositions were then subjected into bitumen tests characterization included penetration test, softening point test, ductility test, flash point test, specific gravity test, solubility and affinity test. The results are summarized in Table 2. Damar asphalt on modification of code S2 selected as the best composition with which the resin powder composition (350 g), gum rosin chunks (100g), fly ash (150 g), and waste cooking oil (205 g).

Table 2. The properties of damar asphalt with various ingredients

No.	Asphalt properties	Unit	Asphalt Specification			Damar asphalt composition and properties		
			Pen 40	Pen 60	Pen 80	S1	**S2**	S3
1	Penetration	dmm	40 - 59	60 - 79	80 - 99	65	**68**	68.4
2	Softening Point	°C	51 - 63	40 - 58	46 - 54	57	**56**	56.25
3	Ductility	Cm	100	100	100	40.5	**63.5**	19.5
4	Flash Point	°C	200	200	225	245	**240**	240
5	Specific Gravity	g/mm	1	1	1	1.003	**1**	0.97
6	Solubility	%	99	99	99	99.84	**95**	78.75
7	Affinity	%	99	99	99	99	**99**	99

3.1.2 Determination of the latex content to improve damar asphalt properties
The polymer of latex is added into the ingredient to achieve the requirement of the ductility test. Table 3 and Figure 1 present the role of latex into the damar asphalt penetration. The effect of latex addition into the ingredient decrease the penetration of damar asphalt.

Table 3. The properties of damar asphalt with various latex content

No.	Asphalt properties	Unit	Asphalt Specification			Latex content (%)				
			Pen 40	Pen 60	Pen 80	0	2	3	4	5
1	Penetration	dmm	40 - 59	60 - 79	80 - 99	68	68.4	59.3	**43**	26
2	Softening Point	°C	51 - 63	40 - 58	46 - 54	56	56.25	57.5	**57.5**	58.5
3	Ductility	Cm	100	100	100	63	19.5	27	**119.5**	114.5
4	Flash Point	°C	200	200	225	250	240	250	**260**	270
5	Specific Gravity	g/mm	1	1	1	0.99	0.97	0.98	**0.97**	0.99
6	Solubility	%	99	99	99	95	78.75	76.25	**95**	99.84
7	Affinity	%	99	99	99	99	99	99	**99**	99

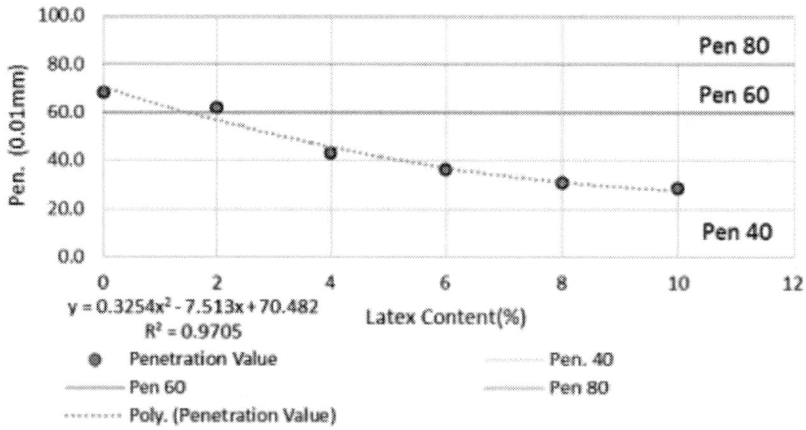

Figure 1. The effect of latex content into the penetration

Figure 2 shows the effect of latex addition into the softening point of damar asphalt. The addition of latex increase the softening point of the ingredient.

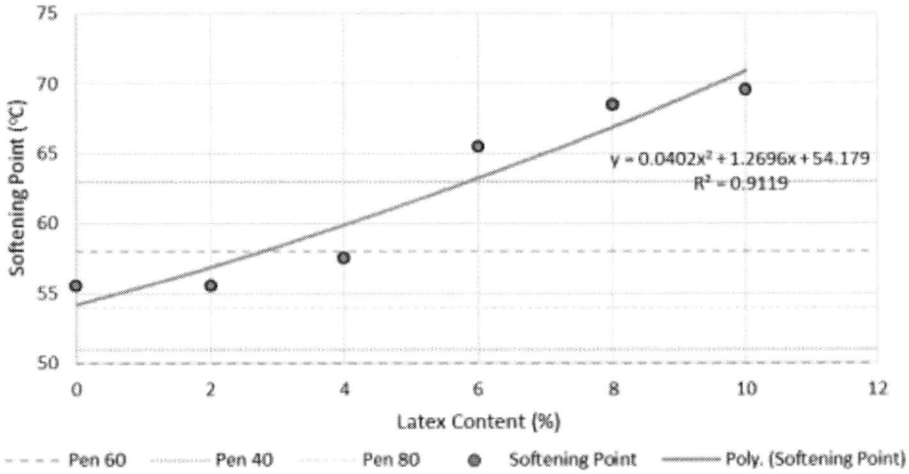

Figure 2. The effect of latex addition into the softening point of damar asphalt

The main objective of latex addition is to increase the ductility test, which is the main problem pure of damar asphalt; it could be seen in Figure 3 that by adding a latex content of 4% the ductility of damar asphalt could be more than the requirements of 100cm.

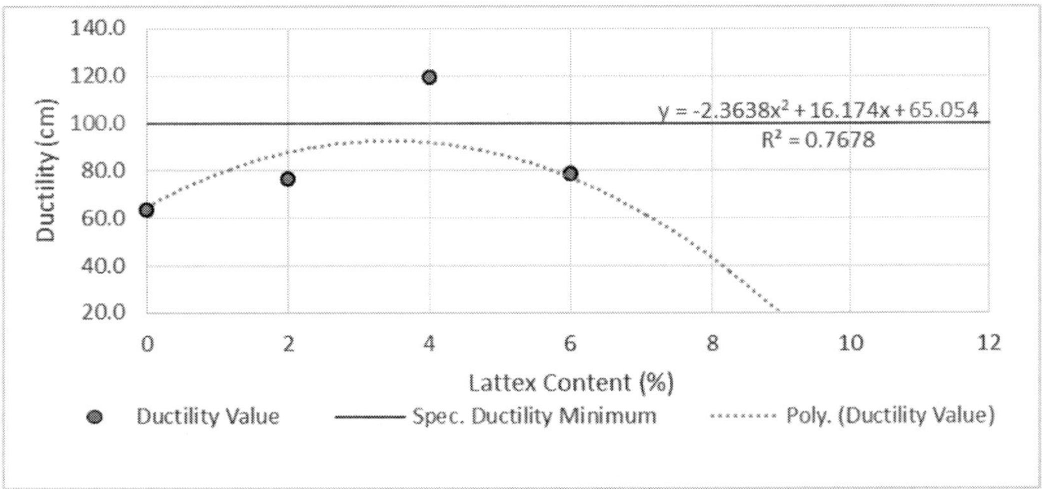

Figure 3. The effect of latex addition into the ductility of damar asphalt

4. Conclusion

Based on the results of testing that was done on damar asphalt, the composition that meets the requirements of penetration bitumen pen 60/70 is a damar asphalt composition with gum rosin (rosin resin powder 350g + mashed 100g), fly ash 0%, cooking oil 205g, but for ductility requirements the composition is not eligible. The most optimal composition obtained in this test is the composition of gum rosin (rosin resin powder 350g + mashed 100g), fly ash 0%, cooking oil 205g with latex content of 4% with a penetration value of 43, the value of the softening point of 57.5°C, value a flash point of 260°C, the value of the burning point 275°C, ductility 119.5cm and a viscosity of 99%. The composition does not meet the requirements for specific gravity of 0.97 and a solubility of only 95%, but due to the penetration of just 43, the most optimal composition is more similar to asphalt pen 40.

References

[1] Wen H, Bhusal S, & Wen B 2012 *J.Mater. Civ.Eng.* **25** 1432-1437

[2] Alvergue A, Elseifi M, Mohammad L N, Cooper Jr S B and Cooper III S 2014 *Road Materials and Pavement Design* **15** 62-77

[1] Baihaqi A, Setyawan A, & Djumari D 2015 *Matriks Teknik Sipil* **3** 4

[2] Nasution M F, Setyawan A, & Sumarsono A, 2015 Proceeding UMS

[3] ACI Committee 226 1987 *ACI Materials Journal Sep-Oct* Michigan, Detroit

[4] Afrianita R 2010 Thesis Andalas University Padang

[6] Departemen Pekerjaan Umum 1991 *SNI 03-1737-1989 Tata Cara Pelaksanaan Lapis Aspal Beton (Laston) Untuk Jalan Raya (*Badan Penelitian dan Pengembangan PU)

[7] Departemen Pekerjaan Umum, 1991 *SNI 06-2456-1991 Metode Pengujian Penetrasi Bahan-Bahan Bitumen* (Badan Penelitian dan Pengembangan PU)

[8] Departemen Pekerjaan Umum, 1991 *SNI 06-2434-1991 Metode Pengujian Titik Lembek Aspal Dan Ter* (Badan Penelitian dan Pengembangan PU)

[9] Departemen Pekerjaan Umum, 1991 *SNI 06-2433-1991 Metode Pengujian Titik Nyala Dan Titik Bakar Dengan Cleve Land Open* Cup (Badan Penelitian dan Pengembangan PU)

[10] Departemen Pekerjaan Umum, 1991 *SNI 06-2488-1991 Metode Pengujian Fraksi Aspal Cair Dengan Cara Penyulingan* (Badan Penelitian dan Pengembangan PU)

[11] Departemen Pekerjaan Umum, 1991 *SNI 06-2432-1991 Metode Pengujian Daktilitas Bahan-Bahan Aspal* (Badan Penelitian dan Pengembangan PU)

[12] Departemen Pekerjaan Umum, 2003. *RSNI S-01-2003Spesifikasi aspal keras berdasarkan penetrasi* (Badan Penelitian dan Pengembangan PU)

[13] Departemen Pekerjaan Umum, 2010. *Rancangan Spesifikasi Umum Bidang Jalan dan Jembatan.* Bandung. Puslitbang Jalan dan Jembatan Badan Penelitian dan Pengembangan.

[14] Namiroh N, 1998 *Pemurnian Damar (*shorea javanica*) Dengan Kombonasi Pelarut Organik,* (Institut Pertanian Bogor: Bogor)

[15] Setianingsih N, 1992 *Pemurnian Damar Shorea javanica Menggunakan Pelarut Organik dan Bahan Pemucat* (Institus Pertanian Bogor: Bogor)

IOP Conf. Series: Earth and Environmental Science 75 (2017) 012020 doi:10.1088/1755-1315/75/1/012020

Environmental Management Model for Road Maintenance Operation Involving Community Participation

A R H Triyono[1], A Setyawan[2], Sobriyah[3] and P Setiono[1]

[1]The student of Ecological Science Doctorate Program of Sebelas Maret University
[2]Roadmate Research Group, Sebelas Maret University
[3]Lecturer of Civil Engineering Department, Sebelas Maret University

E-mail: arysetyawan@staff.uns.ac.id

Abstract. Public expectations of Central Java, which is very high on demand fulfillment, especially road infrastructure as outlined in the number of complaints and community expectations tweeter, Short Mail Massage (SMS), e-mail and public reports from various media, Highways Department of Central Java province requires development model of environmental management in the implementation of a routine way by involving the community in order to fulfill the conditions of a representative, may serve road users safely and comfortably. This study used survey method with SEM analysis and SWOT with Latent Independent Variable (X), namely; Public Participation in the regulation, development, construction and supervision of road (PSM); Public behavior in the utilization of the road (PMJ) Provincial Road Service (PJP); Safety in the Provincial Road (KJP); Integrated Management System (SMT) and latent dependent variable (Y) routine maintenance of the provincial road that is integrated with the environmental management system and involve the participation of the community (MML). The result showed the implementation of routine maintenance of road conditions in Central Java province has yet to implement an environmental management by involving the community; Therefore developed environmental management model with the results of H1: Community Participation (PSM) has positive influence on the Model of Environmental Management (MML); H2: Behavior Society in Jalan Utilization (PMJ) positive effect on Model Environmental Management (MML); H3: Provincial Road Service (PJP) positive effect on Model Environmental Management (MML); H4: Safety in the Provincial Road (KJP) positive effect on Model Environmental Management (MML); H5: Integrated Management System (SMT) has positive influence on the Model of Environmental Management (MML). From the analysis obtained formulation model describing the relationship / influence of the independent variables PSM, PMJ, PJP, KJP, and SMT on the dependent variable MML as follows: MML = 0.13 + 0.07 PSM PJP PMJ + 0.09 + 0.19 + 0.48 KJP SMT + e

1. Introduction

The provincial road organization program in Central Java included road regulation, building, and supervision was organized by the Central Java Province's Office of Highways. Road Organization prioritizes the periodical maintenance, treatment, and examination of road to maintain road service level corresponding to the specified minimum standard service. The availability of funds, human resource, tools and changing legislation became an interesting challenge to be developed more in order to improve

[1] To whom any correspondence should be addressed.

the performance of road [1]. The Central Java Province's Office of Highways as a government institution with main duty and function of managing the road in the improvement, periodical maintenance and routine maintenance program. It has been described that in 2015, the road dealt with by the Office of Highways is 2.565.621 Km in length, 85.09% (2.183.15 Km) with routine maintenance and 14.91% with rehabilitation, periodical maintenance and improvement programs. The bridge dealt with is 26,086.0 M, 97.37 % with routine maintenance program and 2.63 % with rehabilitation, and bridge replacement program [2]. The Javanese people's expectation is very high to the fulfillment of need particularly for road infrastructures included in many of their complaints and expectation in *tweeter, short mail message* (SMS), e-mail, and community's report in various mass media; *Dinas Bina Marga* (Office of Highways) of Central Java Province requires the development of environment management in road routine organization by involving the community participation in the form of fulfilling the representative road condition, that can cater to the road users securely and comfortably [3].

The implementation of road maintenance is highly affected by weather, geographic, and land condition in individual internodes. Recalling that nearly all internodes of provincial road undertake routine maintenance, in which an environmental protection is needed in individual internodes physically, sustainable culture should be created in organization and society, and moral values and mutual trust among the elements of organization in the Central Java's Office of Highways should be implanted, the good environment management should be applied [4].

2. Experimental

This research employed survey method, that was, the research method would use data to generalize the population of research and the data needed in this research included primary and secondary data. This research started by specifying a model based on the theory. Individual variables in the model were conceptualized as the latent ones and measured with some indicators. Some indicators were developed for individual models. The factor analysis was used to determine the indicators to be used to measure the relevant latent variables and represented by some factors [5,6].

The collected and qualified data with adequate validity and reliability was analyzed using the tabulation calculation system corresponding to individual variables used in this research. Data processing was carried out using Structural Equation Model (SEM) based on Variance and Component-Based SEM with Smart PLS software [7].

3. Results and Discussion

The research variable is the measurable abstract concept (observed variable). However, there is also the immeasurable abstract concept (unobserved variable). This dissertation research is intended to examine the multidimensionality of construct consisting of six variables.

3.1 Latent Independent Variable

a.The first latent independent variable is Community Participation in governing, building, developing and supervising road (PSM) (X1), constituting the variable affecting other variable, and in this research, the latent independent variable consisting of 5 indicator variables (X1-1,X1-2, X1-3, X1-4, X1-5): X1-1 is the Level of Community participation in the activities in road area; X1-2 is the form of Community participation in the activities in road area, X1-3 is the Community's need for road function, X1-4 is the community's readiness in active participation; and X1-5 is local wisdom.

b.The second latent independent variable is Community behavior in utilizing road (PMJ) (X2), constituting the variable affecting other variables, and in this research the latent independent variable consisting of 5 indicator variables (X2-1,X2-2, X2-3, X2-4, X2-5): X2-1 is the Community's knowledge on legislation pertaining to road, X2-2 is the Community's compliance with the legislation pertaining to road, X2-3 is the Discipline of maintaining the road performance, X2-4 is the Community's compliance with the payment of road retribution, and X2-5 is the care about the problems pertaining to the road surrounding.

c.The third latent independent variable is the Provincial Road Service (PJP) (X3) is the variable affecting other variables, and in this research the latent independent variable consisting of 5 indicator variables X3-1,X3-2, X3-3, X3-4, X3-5): X3-1 is The development of provincial road maintenance, X3-2 is the implementation of routinely provincial road maintenance, X3-3 is the supervision over provincial road maintenance, X3-4 is the quick handling of provincial road damage, and X3-5 is the evaluation on the sustainable implementation.
d.The fourth latent independent variable is work safety in Provincial road (KJP) (X4), constituting the variable affecting other variable, and in this research the latent independent variable consisting of 5 indicator variables (X4-1,X4-2, X4-3, X4-4, X4-5): X4-1 is the readiness of wok security cost, X4-2 is the management plan in work safety, X4-3 is the availability of personnel in supporting work safety, X4-4 is the availability of work safety-supporting tool, and X4-5 is the follow-up of work safety program.
e.The fifth latent independent variable is an Integrated Management Sistem (SMT) (X5), constituting the variable affecting other variable, and in this research the latent independent variable consisting of 5 indicator variables (X5-1,X5-2, X5-3, X5-4, X5-5): X5-1 is the study of planning, X5-2 is the study of implementation, X5-3 is the study of examination and evaluation, X5-4 is the study of improvement measure, and X5-5 is the study of SKPD's commitment.

3.2 Latent Dependent Variable (Y)
The variable affected by other research in this research was a routinely provincial road maintenance, integrated into an environment management system and involving community participation.

3.3 Convergent Validity Test
This convergent validity test was intended to test whether or not the indicators used is fairly valid and representative in measuring its construct. The criteria used in this validity test:

a. P value < 0.05
b. Loading factor> 0.7 (valid);
c. Loading factor = 0.4 s/d <0.7 (remaining to be maintained as long as it does not raise the AVE value and the composite value is reliable above its threshold);
d. Loading factor < 0.4 (the indicator should be dropped out of the model).

From the SEM PLS analysis, the factor loading value result was obtained of 0.4 for each of the variables, so that those indicators are fairly relevant in measuring PSM, PMJ, PJP, KJP, SMT and MML

3.4 Discriminant Validity Test
This validity test was intended to examine whether or not the indicators used are fairly valid in contributing to R^2. The criteria used in this validity test are: root value AVE > correlation value between constructs in the same column:

a. Root value of AVE for PSM variable: 0.642 > correlation values between constructs in the same column.
b. Root value of AVE for PMJ variable: 0.668 > correlation values between constructs in the same column.
c. Root value of AVE for PJP variable: 0.688 > correlation values between constructs in the same column.
d. Root value of AVE for KJP variable: 0.683 > correlational values between constructs in the same column.
e. Root value of AVE for SMT variable: 0.717 > correlational values between constructs in the same column.
f. Root value of AVE for MML variable: 0.703 > correlation values between constructs in the same column.

Thus the indicators of variable belong to *valid* category in discriminant way.

3.5 Construct Reliability Test

This test is intended to test the reliability of the instrument/indicator creating a construct. The criteria used in this test are:

a. Composite reliability > 0,7
b. Cronbach's alpha > 0.7

The composite reliability and cronbach's alpha values can be seen in the Table 1.

Table1. Latent Variable Coefficient

Composite reliability coefficients					
PSM	PMJ	PJP	KJP	SMT	MML
0.844	0.863	0.863	0.897	0.927	0.886
Cronbach's alpha coefficients					
PSM	PMJ	PJP	KJP	SMT	MML
0.786	0.816	0.814	0.872	0.914	0.853

From the table above, it can be interpreted as follows:

a. The composite reliability values for PSM, PMJ, PJP, KJP, SMT and MML constructs are: 0.863; 0.863; 0.897; 0.927 and 0.886, respectively, all of which are higher than 0.7;
b. The cronbach's alpha values for PSM, PMJ, PJP, KJP, SMT and MML constructs are: 0.816; 0.814; 0.872; 0.914 and 0.853, respectively, all of which are higher than 0.7. Thus it can be concluded that all variables in this research ((PSM, PMJ, PJP, KJP, SMT and MML) are reliable.

3.6 Hypothesis Testing

The hypothesis to be tested includes:

a. H1: Community Participation (PSM) affects positively the Environment Management Model (MML)
b. H2: Community behavior in utilizing road (PMJ) affects positively the Environment Management Model (MML)
c. H3: Provincial Road Service (PJP) affects positively the Environment Management Model (MML)
d. H4: Work Safety in Provincial Road (KJP) affects positively the Environment Management Model (MML)
e. H5: Integrated Management System (SMT) affects positively the Environment Management Model (MML)

From the analysis using SEM-PLS 3.0, the path analysis diagram is obtained as follows:

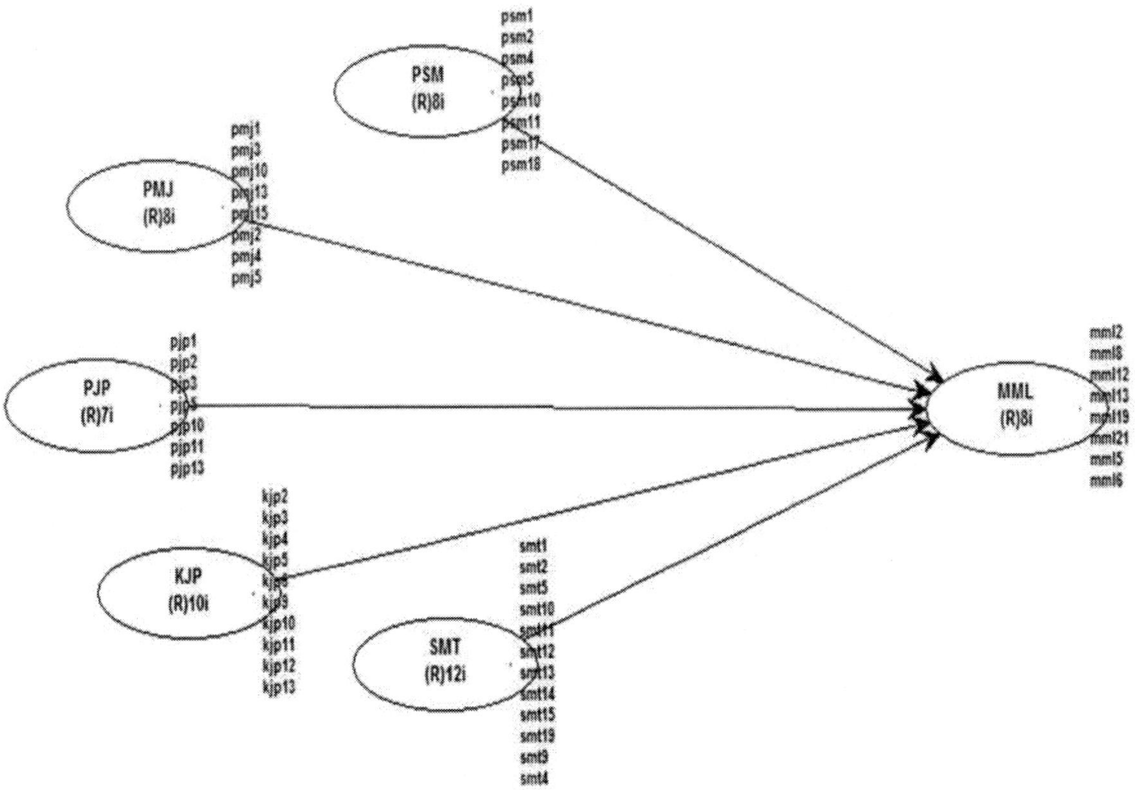

Figure 1. Path diagram (without parameter value)

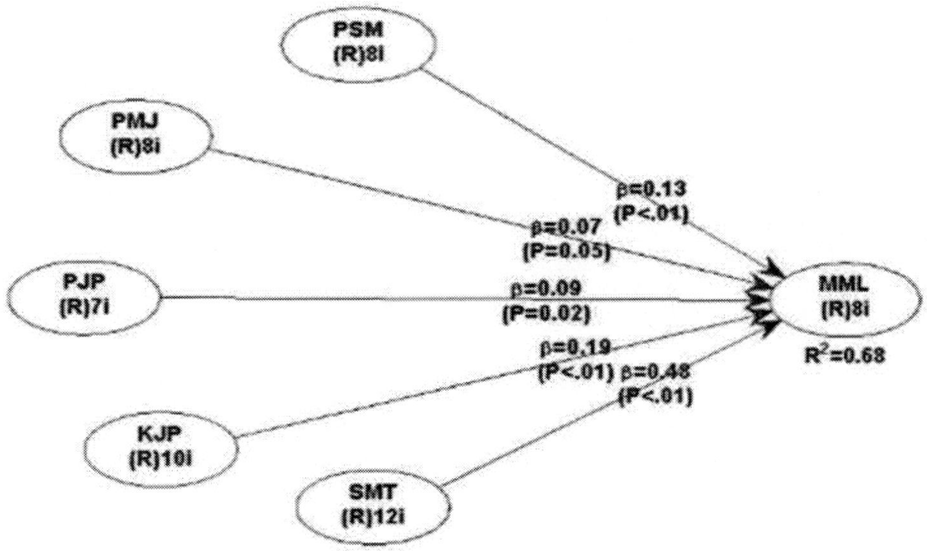

Figure 2. SEM Analysis Completed (with its variable values)

Model Fit Indices and P Value
APC=0.194. P<0.001

ARS=0.760. P<0.001
AVIF=2.728. Good Ir< 5
From the chart and model fit indices and value, it can be interpreted as follows:

a) Model and coefficient values contained in the chart is fit (APC = 0.194 with p < 0.001 , ARS = 0.68 and p < 0.001 and AVIF = 2.299 < 5);
b) Predictor variable variation (PSM, PMJ, PJP, KJP, SMT) explains the 68% of criterion variable variation MML, including in the criterion of explaining overall (large);
c) Hypothesis 1 stating that PSM affects positively MML is supported (p=0.01< 0.05)
d) Hypothesis 2 stating that PMJ affects positively MML is supported (p=0.02< 0.05)
e) Hypothesis 3 stating that PJP affects positively MML is supported (p=0.02< 0.05)
f) Hypothesis 4 stating that KJP affects positively MML is supported (p=0.01< 0.05)
g) Hypothesis 5 stating that SMT affects positively MML is supported (p=0.01< 0.05)

4. Conclusion

1. There is an environment management model in organizing the routinely provincial maintenance by involving community participation obtained using SEM analysis.
2. From the analysis, the model formulation is obtained describing the relationship/effect of independent variables, including PSM, PMJ, PJP, KJP, and SMT on the dependent one, MML, as explained below:

MML = 0.13 PSM + 0.07 PMJ + 0.09 PJP + 0.19 KJP + 0.48 SMT + e

References

[1] Anonymous 1996 *Pedoman Teknis Kajian Aspek Sosial dalam Penyusunan Analisis Mengenai Dampak Lingkungan* (Jakarta: Keputusan Kepala Badan Pengendalian Dampak Lingkungan nomor KEP-299/11/1996)
[2] Anonymous 2012 *Pedoman peran serta masyarakat dalam penyelenggaraan jalan* (Jakarta: Peraturan Menteri Pekerjaan Umum Nomor 01/PRT/M/2012)
[3] Arnstein R 1969 *A Ladder of Citizen* Participation in. *Classic Readesign Urban Planning. An Introduction* Jay M. Stein (ed) (New York: McGraw-Hill Inc)
[4] Fondation Environmental Conservation 2010 *Reflecting on the next generation of models for community-based natural resources management*
[5] Gaspersz V 2012 *Three-in-One ISO 9001,ISO 14001, OHSAS 18001, Sistem Manajemen Kualitas, K3, Lingkungan (SMK4L) dan Peningkatan Kinerja Terus-Menerus* Jakarta: Vinchristo Publication)
[6] Warren D at al 2005 *Linking Community Empowerment, Decentralized Governance, and Public Service Provision Trought a Local Development Framework*
[7] Ghozali I 2008 *Structural Equation Modeling Metode Alternatif dengan Partial Least Square PLS 2nd Edition* (Semarang: UNDIP Publishing)

IOP Conf. Series: Earth and Environmental Science **75** (2017) 012021 doi:10.1088/1755-1315/75/1/012021

Microscopy of Alloy Formation on Arc Plasma Sintered Oxide Dispersion Strengthen (ODS) Steel

B Bandriyana[1], A Sujatno, R Salam, A Dimyati, and P Untoro
[1]Center for Science and Technology for Advanced Materials, BATAN-Indonesia

E-mail: bandri@batan.go.id

Abstract. The oxide dispersed strengthened (ODS) alloys steel developed as structure material for nuclear power plants (NPP) has good resistant against creep due to their unique microstructure. Microscopy investigation on the microstructure formation during alloying process especially at the early stages was carried out to study the correlation between structure and property of ODS alloys. This was possible thanks to the arc plasma sintering (APS) device which can simulate the time dependent alloying processes. The ODS sample with composition of 88 wt.% Fe and 12 wt.% Cr powder dispersed with 1 wt.% ZrO2 nano powder was mixed in a high energy milling, isostatic compressed to form sample coins and then alloyed in APS. The Scanning Electron Microscope (SEM) with X-ray Diffraction Spectroscopy (EDX) line scan and mapping was used to characterize the microstructure and elemental composition distribution of the samples. The alloying process with unification of each Fe and Cr phase continued by the alloying formation of Fe-Cr by inter-diffusion of both Fe and Cr and followed by the improvement of the mechanical properties of hardness.

1. Introduction.

Research on special high temperature materials plays an important role in supporting the national nuclear energy program. Meanwhile due to its superior properties at high temperature, high chromium oxide dispersed strengthened (ODS) alloys are believed to be the best candidate as structure material for nuclear power plant (NPP) of 4th generation [1,2]. Their origin can be found especially on their good resistant against oxidation and creep at elevated temperature. This is because of their unique microstructure that formed during alloying process. However, in this case the early stages of phase formation must play a critical role. Therefore, the mechanism of alloying process of the ODS steel was necessary investigated to get reliably and accurate information that is important for improving the performance of the alloy manufacturing. Commercial ODS steels are usually added by the Yttrium oxide Y2O3 particles up to 1 wt% [3]. The zirconia particles are, however, more proposed for application as nuclear structure material because of their low neutron absorption cross section. On the other hand, the addition of zirconium to a ferritic matrix of ODS steel could also effectively improve the performance against high temperature and corrosion by the formation of more stable Y-Zr oxide [4,5]. Manufacturing of ODS steels was normally performed via powder metallurgical route which can easily disperse the oxide particles homogeneously. However, this process is limited by the amount of material can be added, costly and very time consuming [6]. Therefore, many researches was worldwide conducted in developing of new manufacturing technique which all are aimed in finding of processes that are faster and lower in cost. A casting route will be the most effective in accordance to the cost and production capacity, but it exhibits difficulties in uniform dispersion of the ceramic particles in liquid metal for the chemical reactions with the matrix elements [7]. Recent method using Spark Plasma Sintering (SPS) can obviously produce ODS alloy with quite good quality by relatively fast process time, but this technology still need high process energy [8]. In this work the proposed new method of Arc Plasma Sintering (APS) which uses plasma as heat source has been developed in PSTBM-BATAN can significantly reduce process time and energy consumption. This apparatus is used for synthesizing of alloys with different composition. Recent work reported the fabrication of Fe-Cr and feritic ODS alloys to obtain the optimal properties on the high temperature material application [9]. In this work, microscopy investigation on the alloying process especially at the first time of their formation during ODS steel production was performed with help of arc plasma sintering (APS) with the aim to study the possible correlation between structure and property of ODS alloys.

2. Experimental

For the experiment the samples composed of 88 wt.% Fe and 12 wt.% Cr powder with addition of 0.5 and 1 wt.% ZrO2 nano powder as dispersoids were prepared. The powder were mixed in a rotary mill for 3 hours and were isostatic pressed with 20 tons of load producing sample coins of 15 mm of diameter and 4 mm thick. The sample coins were then consolidated in APS at 12 V and 25 Amps with time variation ranges between 2 and 4 minutes. The technical drawing of the experimental set up in APS is shown schematically in Fig. 1. In this work the F-Cr ODS ferritic steel was developed with 12 wt.% Cr and 0.5 and 1 wt.% of the dispersion particles ZrO_2. The samples were prepared by mixing and milling all the materials (Fe, Cr and ZrO_2 powder) for 1, 2 and 3 hours. After milling the samples were pressed by isostatic compress machine with the compression load of 20 Ton to form sample coins. The samples were then consolidated by sintering in the APS for 4 minutes. The principal design of the APS is illustrated schematically in Fig. 1 [6]. The process was carried out in fully argon atmosphere in order to protect against oxidation. The sample coin was placed on a copper cup and exposed by the plasma with the power 300 Watt and variation of time.

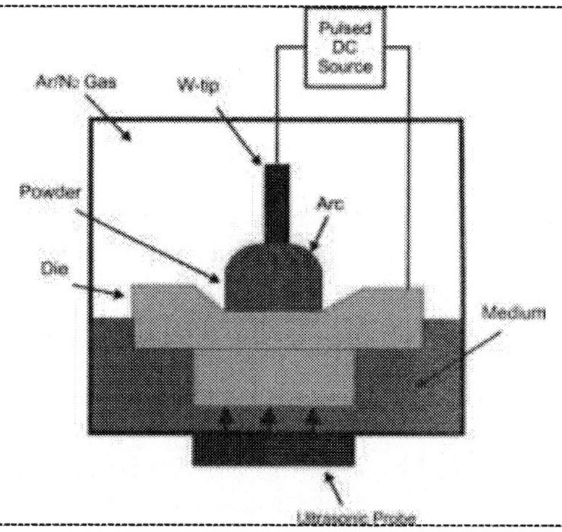

Figure 1. Schematic diagram of APS

APS is an experimental sintering and melting device newly developed at the Centre of Science and Technology for Advanced Materials (PSTBM) of Indonesia nuclear energy agency [10]. Its heart is an arc plasma source, which can provide a very high process temperature in a short time. This allows a fast process of densification or consolidation of almost every powder metallurgical prepared samples. Therefore, APS is suitable for simulation of some thermal induced processes such as alloy formation. The Scanning Electron Microscope (SEM) equipped with X-ray Diffraction Spectroscopy (EDX) was used to characterize the microstructure and elemental composition of the samples. EDX line scan and area mapping were applied in order to visualize the elemental distribution in the sample. The mechanical properties of the samples were measured by means of Vickers micro hardness.

3. Results and Discussion

Figure 2 shows the SEM image of the sample after compression 20 tons. One can see clearly every single Fe and Cr particle. The result of the press process is quite good. The amount of porosity is about less than 15%.

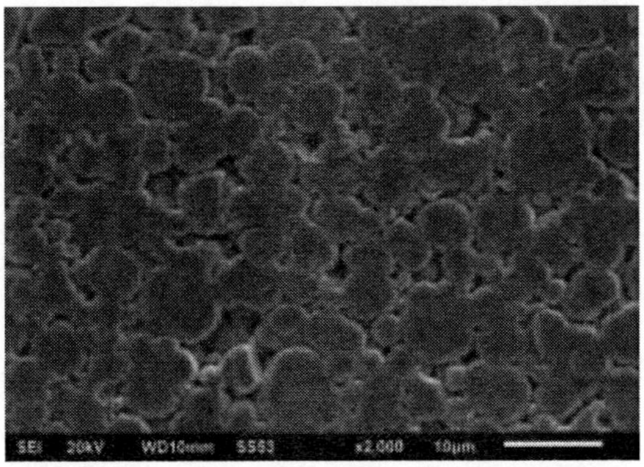

Figure 2. SEM image of the sample after compression 20 tons

The microstructure of ODS ferritic alloys dispersed by 0.5 and 1 wt.% were observed by SEM. The results are illustrated in Figure 2. The SEM micrograph in Figure 3a shows general appearance of the sample microstructure after 4 minutes consolidation through sintering in APS taken by SEM at 20 kV using backscattered electrons detector (BSE). The sample consists of two phases, the white matrix phase and large precipitate phase. The porosities have been completely disappears. From the EDX analysis it is obvious that the matrix (zone 1) mainly consists of iron phase with 96.68 wt.% Fe and 1.60% Cr. While, the zone 2 is chromium phase dominated by 96.51 % Cr and Fe 1.85 %. This is assumed to be caused by the similarity of Gibbs free energy of formation of single elements compared to those of two and more elements. The EDX line scan in Figure 3b was put along the straight line as indicated in Figure 3a showing the compositional changes of the Fe and Cr in the phase interface. It can be observed that the distribution graph of both Fe and Cr exhibits characteristic diffusion zone, which elucidate the diffusion in two directions. Fe and Cr are moving into each other. However, the concave flow of the graph explains that the Cr diffusion must be faster than Fe. From the result can be concluded that the formation of Fe-Cr alloy begins by the melting of each alloy element forming the Fe and Cr phases then continued with the formation of Fe-Cr alloy which is determined by the diffusion process. This diffusion process is possible because of the existence of the relatively high chemical potential due long range elemental in homogeneity in the Fe matrix and Cr phase.

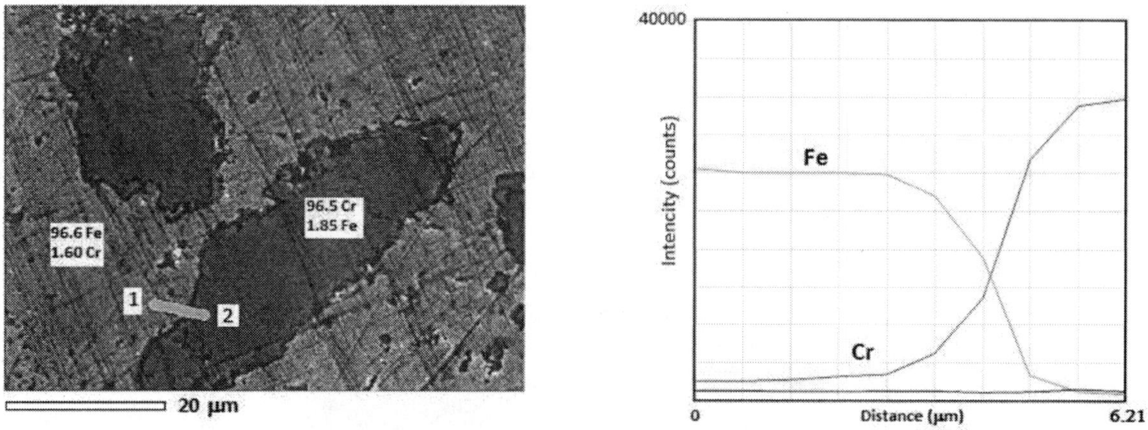

a. **b.**

Figure 3. SEM image of ODS steel produced by APS (a) line scan mapping (b)

The behavior of the alloy sintered by APS differ significantly to the results of conventional sintering using the sintering furnace as shown in Figure 4.

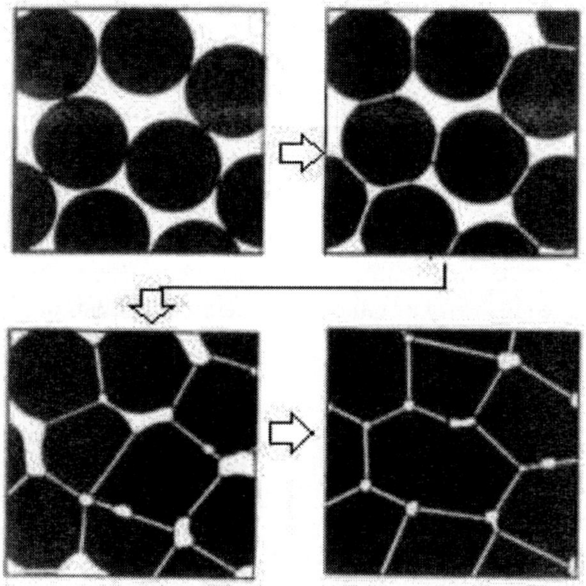

Figure 4. Alloying scheme of conventional sintering process

The early portion of sintering produce bonds between contacting particle and focuses on the growth of interparticle bonds, followed by densification and pore elimination. The process was continued by the diffusion processes at sintering temperature that cause necks to form and grow the grain. The zirconia was finely dispersed in the matrix and in the grain boundaries of both Fe and Cr phase as well. From the observation on all the samples it is revealed that the alloying process was strongly depend on the sintering time. However, after 4 minutes sintering showed the more proper alloy of Fe with Cr dispersed phase. The hardness of 150 VHN after 4 minutes of sintering was achieved showing the improvement of the mechanical properties caused by the alloying process during sintering process.

4. Conclusion

An experimental study on the alloy formation during early stages of their formation using APS has been performed. The alloying process begins with unification of each Fe and Cr phase due to preferred low Gibbs free energy for the formation of phase of the same atoms. With increasing time, the development of Fe-Cr alloy formation in dependency on sintering time. Alloy formation is the result of inter-diffusion of both Fe and Cr in both directions forming zone of Fe-Cr phase, however Cr moves faster into the new formed phase than Fe. These development followed by the improvement of the mechanical properties that can be observed on the hardness of 150 VHN after 4 minutes of sintering.

Acknowledgment

The authors thank to Ministry of Research, Technology and Higher Education (Kemenristekdikti) for financial support under the INSINas 2016 program, RT2016-0287.

References

[1] Klueh R L, Shingledecker J P R, Swindeman R W, and Hoelzer D T 2005 *J. Nucl. Mater.*
[2] Chuto T, Nagase F, Fuketa T 2009 *Nucl. Eng.Tech.* **41** 2
[3] Subramanian K, Angelo P C, and Ramanathan S 2010 *Trans. Indian Inst. Met.* **63** 67-74
[4] Garcia N, Rodriguez 2015 *J. Am. Ceram. Soc.*
[5] Gao R, Tao Zhang, Wang X P, Liu C S 2014 *J. Nucl. Mater.*
[6] Raman L, Gothandapani K, Murty B S 2016 J. *Defence Science* **66** 316-322
[7] Shi Z , Haan F 2015 *Mater. Res. Innovations* **19**
[8] Zhang H, Huang Y, Ning H, Williams A 2015 *J. Nucl. Mater.* **464** 61–68
[9] Dimyati A 2016 *Proc. Asia-Pacific Microscopy Conference 11*
[10] Dimyati A 2015 *Technical Report of Insentif Riset Sistem Inovasi Nasional*

Flat Plate Solar Collector Characteristic with Shutter Glass Distance Variation and Collector Inclination Angle

A H Ramelan[1], A Burhanuddin[1], M I A Fuady[1], S Wahyuningsih[1], H Munawaroh[1]

[1]Research Center for Material Electronic and Energy, Sebelas Maret University, Indonesia

E-mail: aramelan@mipa.uns.ac.id

Abstract. It has been done a research to determine the solar thermal collector efficiency of the flat plate collector. The testing of the collector was conducted on 30th November 2005, 1st, 3rd, 5th, 6th, 7th December 2005 with the variation of distance between one glass covers were varied from 3 cm, 6 cm, and 9 cm; and with inclination angle variation of collector from 10°, 20°, 30°, and 40°. Solar thermal collectors absorb the radiant energy from the sun and convert it to heat between the bottom glass cover and absorbing plates in the collector. Parameters which influence on the collector performance include distance between plate collector with glass covers and the inclination angle. It was found that the difference between output- input temperatures is the highest on a distance of 3 cm and inclination angle of 10°. This is influenced that inclination angle 10° more close to zenith angle. The solar thermal collector efficiency is not a constant; The solar collector efficiency depends on solar radiation intensity, input-output temperature difference and air flow. The smaller the inclination angle of solar collector, the higher the absorption radiation. If inclination angle of collector same with zenith angle, so the absorption radiation will maximum.

1. Introduction

Indonesia has a tropical climate which is relatively high ambient temperature, relative humidity, and in some places have high rainfall as well. Indonesia is also known as an agricultural country that produces in addition to staple foods also produce other agricultural products such as cocoa, coffee, copra, nutmeg and others [1]. Commodities were mostly should be dried immediately after harvest, because if too late there will be a process of decay so very detrimental.

To dry it takes enormous energy. Farmers mostly do the drying in the hot sun. Ambient temperature is about 33 °C, while the drying temperature for late-many agricultural commodities ranging from 60-70 °C. If we use the air heater temperature environment or lower than the drying temperature, it will take longer. Raising the temperature of the environment can be done by collecting the air in a solar collector and exhale into commodities [2].

Fossil energy, especially petroleum, is the main energy source and limited in number. The limited resources of fossil energy cause the need for renewable energy development. Renewable energy is a non-fossil fuels derived from natural and renewable [3]. When properly managed, these resources will not be exhausted.

Indonesia, on the one hand is an archipelago so that the transportation of commercial energy will remain an obstacle to the provision of cheap energy in remote places mentioned above. On the other hand, Indonesia has the potential of renewable energy sources is large enough [4]. In the future, the potential for developing renewable energy sources and has a great chance given the strategic nature of renewable energy sources is a source of clean energy, environmentally friendly and sustainable.

1.1 Solar energy and utilization

As the nearest star on the blue planet Earth, which is only about 150 million km, it is natural that only radiant solar energy that affect the dynamics of the atmosphere and life on Earth. The energy that comes to Earth is largely the sun's radiation. This energy is then transformed into various forms of energy, for example warming the Earth's surface, the motion and the warming of the atmosphere, ocean waves, photosynthesis of plants and other photochemical reactions [5].

Spreading sunshine each year varies hemisphere. Indonesia receives an average of eight hours a day of sunlight and the intensity of sunlight entering the specified position of the sun against the collector.

1.2 Review of heat transfer

1.2.1 Conduction

Heat flows by conduction from high temperature areas to temperate areas. Conduction heat transfer rate can be expressed by the Fourier law as follows [6]:

$$q = -kA\left(\frac{dT}{dx}\right) \quad (1)$$

where q is the rate of heat transfer, W; k is the thermal conductivity, W/(m.K); A is the cross-sectional area perpendicular to the flow of hot m^2 and dT/dx is the temperature gradient in the direction of heat flow, -K/m.

1.2.2 Convection

The air flowing over a metal surface on a solar air heater was heated by convection. There are two types of convection process that forced convection and natural convection. The heat transfer rate can be expressed by the following equation [6]:

$$q = hA(T_w - T_f) \quad (2)$$

where q is the rate of heat transfer, W; h is the convection coefficient, W/(m^2.K); A is the surface area, m^2; T_w is the temperature of the walls; and is the temperature of the fluid, K.

For solar heaters that work in the area of Reynolds number between 2000 and 10000, and the value of Nusselt number [5]:

$$N_u = 0.00269.\,Re \quad (3)$$

Re is the Reynolds number that usually ranges from 2000 to 10000 for turbulent flow, and below 2000 for laminar flow. Reynolds number can be formulated [6]:

$$R_e = \frac{vd_i\rho}{\mu} \quad (4)$$

where R_e is the Reynolds number, v is velocity - average of the fluid (m/s), d_i is the diameter of the pipe (m), ρ is the density (kg/m^3), μ is the dynamic viscosity (kg/m.s).

1.2.3 Radiation

Total radiation heat transfer from a perfect black body is proportional to the fourth power of the temperature of the object. This is the Stefan-Boltzmann law so that it can be written as follows [7]:

$$E = \sigma A T^4 \tag{5}$$

where σ is the Stefan-Boltzmann constant magnitude 5.67×10^{-8} W/m^2.K^4, A is the cross-sectional area of the object (m^2), T is the absolute temperature of the object (K).

1.3 Overview fluid mechanics

Viscosity is a trait that determines the characteristics of the fluid which is a measure of the shear fluid prisoners. Dynamic viscosity is defined as the ratio between shear stress and shear strain rate.

1.4 Position of the sun

1.4.1 Equation for Angle Zenit

Θ_z zenith angle is shown as the angle between the zenith z, or a straight line above the head, and the line of sight to the sun. The equation for the zenith angle can be formulated [6]:

$$\cos\theta_z = \sin\delta\sin\phi + \cos\delta\cos\phi\cos\omega \tag{6}$$

where Y_z is the zenith angle, δ is declination, ϕ is the latitude angle, and ω is the hour angle (15° per hour). Declination δ, is the angle formed by the sun in the equatorial plane, it changed as a result of the tilt of the earth, of +23.45° in summer (June 21) and -23.45° in winter (December 21). Declination at any time can be estimated from the following equation [6]:

$$\delta = 23.45\sin\left(360 \times \frac{284 - n}{365}\right) \tag{7}$$

where n is the day of the year.

1.4.2 Intensity radiation on the inclined plane

Component of radiation on a tilted surface, which spotlight I_{bT} component is obtained by converting the radiation beam on the horizontal surface into a normal entry by using the zenith angle, and then get the component on an inclined surface by using a corner entrance

I_{bT} beam radiation on an inclined surface can be calculated from the radiation beam (measurable) I on a horizontal surface [6].

$$I_{bT} = I\frac{\sin\delta\sin(\phi-\beta) + \cos\delta\cos(\phi-\beta)\cos\omega}{\sin\delta\sin\phi + \cos\delta\cos\phi\cos\omega} \tag{8}$$

1.5 Variety solar thermal collectors

1. Flat plate solar collectors
2. Collectors concentrated
3. Tube collector
4. Passive collectors

1.6 Analysis of the working solar thermal collectors flat type

1.6.1 Energy equation equilibrium

Equilibrium equations rate of heat energy in thermal collectors can be expressed by the equation:

$$q_u = q_i - q_l \tag{9}$$

Where q_u is the energy used (J/s), q_i is the incoming energy (J/s) and q_L is the lost energy (J/s).

a. The rate heat energy entering

The rate of heat energy entering the thermal collector solar energy (J/s) affected by the I_{bT} amount of the intensity of solar radiation on an inclined surface (watt/m^2), A_p wide plate absorber thermal collector (m^2), and the product of transmissivity glass cover-absorptivity plate absorber ($\tau.\alpha$), expressed by the equation:

$$q_i = A_p . I_{bT} . (\tau.\alpha) \tag{10}$$

b. The rate heat energy used

The rate of heat energy out of the solar thermal collectors can be expressed in the equation:

$$q_u = \overset{0}{m} . C_p . (T_O - T_I) \tag{11}$$

c. The rate heat energy missing

Not all of the incoming heat energy can be used entirely because there are heat loss factors on thermal collectors. The heat losses occur at the top of the solar thermal collector called heat losses at the top and bottom of the solar thermal collector called the lower heat losses. Whereas, the amount of heat loss is the total heat loss.

- The rate of heat energy losses Section differences (top loss) q_{tl}
- The rate of heat energy losses Bottom (bottom loss) q_{bl}

1.6.2 Efficiency solar collector

The definition of the efficiency of the solar collector is the ratio between the energy used by the amount of solar energy received at a particular time by the solar collectors.

$$\eta = \frac{\dot{m} C_p (T_0 - T_i)}{A_p (\tau\alpha) I_{bT}} \tag{12}$$

air mass flow rate is the amount of air mass that flows per unit time and can be expressed as follows:

$$\dot{m} = \frac{m}{t_u} = \frac{V\rho}{t_u} \tag{13}$$

2. Experimental

2.1 Place and time collection

This study was conducted in a central laboratory backyard UNS Surakarta on November 30 to December 7, at 10:00 to 14:00.

2.2 Tools

All tool were used are thermocouples, anemometer testo, digital thermometers, light Meter Model Li-250 No Sri LMA – 2706, and sensor pyranometer No seri PY – 46415.

IOP Conf. Series: Earth and Environmental Science **75** (2017) 012022 doi:10.1088/1755-1315/75/1/012022

2.3 Research procedure

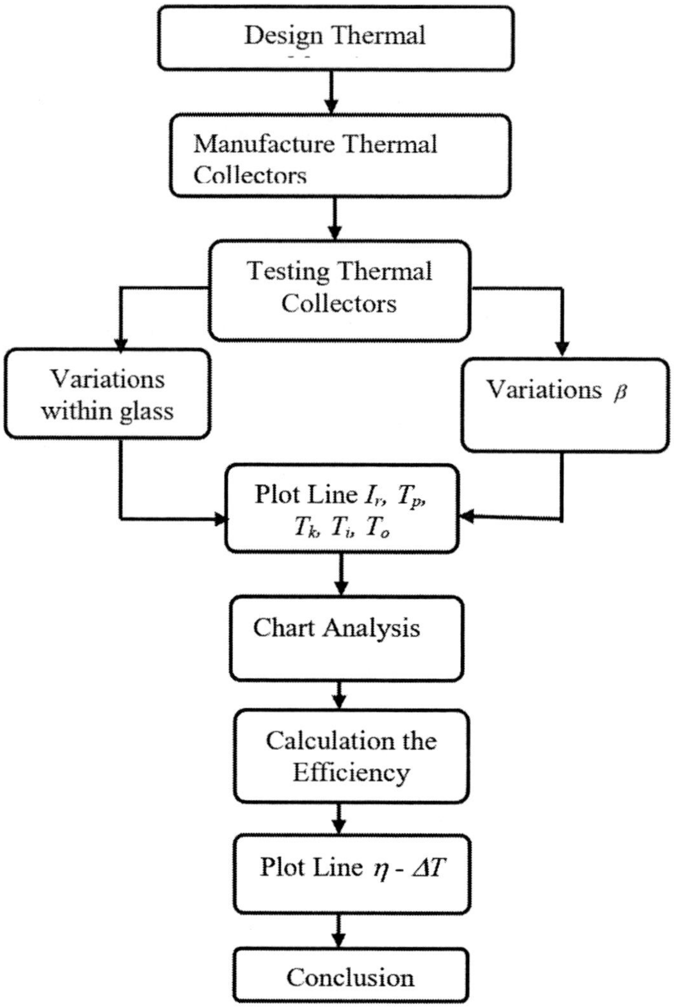

3. Results and Discussion

3.1 Intensity of solar radiation

3.1.1 Temperature variation of the distance collector on glass cover

The results of temperature measurement input and output temperature on the research conducted on December 1, at a distance of 3 cm and 9 cm can be seen in Figure 1 and Figure 2.

In Figure 1 and Figure 2 we can see that a large output temperature greater than the temperature input. At a distance of 3 cm difference plate temperature value, input and output reached 23.1 °C and 9.9 °C little difference. Temperatures highest output reached 63.8 °C at 11:45 and the intake temperature reached 41.7 °C at 12.45. At a distance of 9 cm difference plate input and output value reached 13 °C largest and the smallest difference of 0.4 °C. Temperatures highest output reached 51.2 °C at 11.45 and the intake temperature reached 42.6 °C at 11.45.

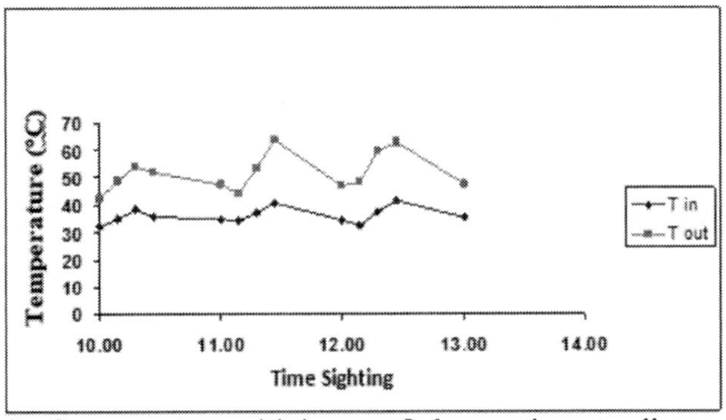

Figure 1. Graph of temperature with hours of observation at a distance of 3 cm

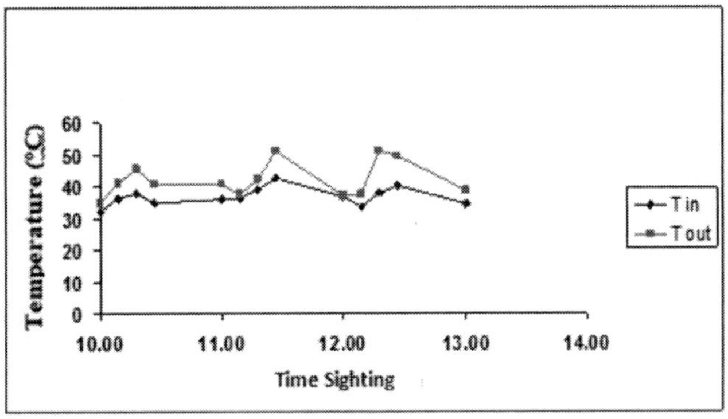

Figure 2. Graph of temperature with hours of observation at a distance of 9 cm

Results temperature solar collectors can be seen in the chart entry of the temperature difference (T_{in}) and the discharge temperature (T_{out}) of the hours of observation :

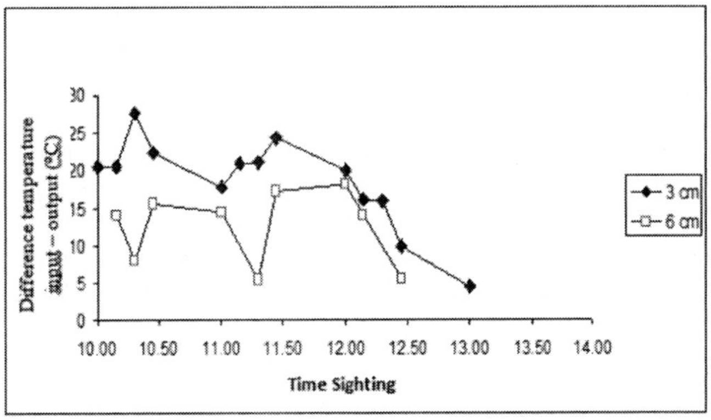

Figure 3. Chart input-output temperature difference at a distance of 3 cm and 6 cm

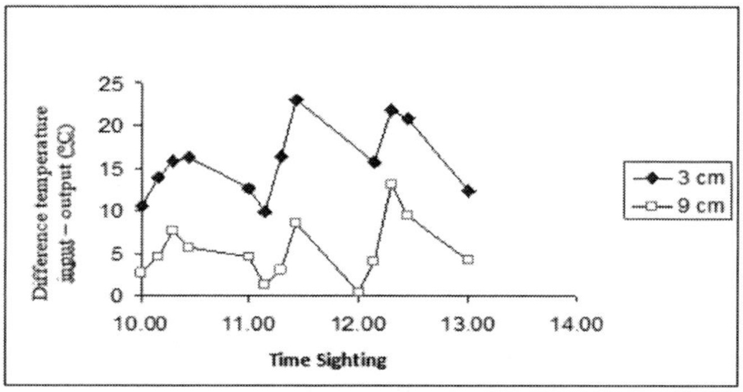

Figure 4. Chart input-output temperature difference at a distance of 3 cm and 9 cm

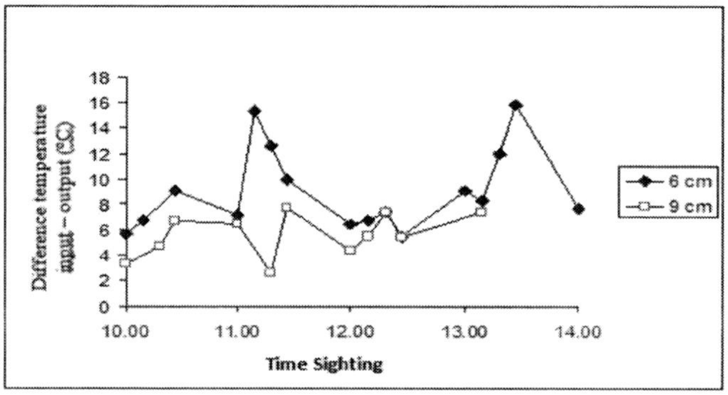

Figure 5. Chart input-output temperature difference at a distance of 6 cm and 9 cm

In Figure 3, Figure 4 and Figure 5, we can see that on November 30, 2005 the difference in temperature at a distance of 3 cm glass the result is higher than the glass 6 cm distance. But there are two data results are the opposite, this is because any change in the flow that moves around the collector. On December 1, 2005, the difference in temperature at a distance of 3 cm glass the result is higher than the glass 9 cm distance. This is because at a distance of 9 cm glass heat is lost to the greater environment. So that heat absorption on the plate is reduced. But on December 3, the difference in temperature at a distance of 6 cm glass results is mostly higher than the glass 9 cm distance. At a distance of 9 cm glass much heat is lost to the environment. But there are some circumstances where the amount of the opposite, this is due to differences in the airflow moving around collector. Base on the measurement results can be concluded that the variation within the glass effect on the temperature difference between the collector. The temperature difference will be maximum at a distance of a little glass, because little heat energy is lost to the environment.

3.1.2 Temperature collector on variation of the angle of the collector

The results of the temperature in the solar collectors can be seen in the chart entry of the temperature difference (T_{in}) and the discharge temperature (T_{out}) of the hours of observation. In Figure 6, Figure 7 and Figure 8 we can see that on December 5, there is a large difference in temperature at a 10° angle no greater than the temperature difference at an angle of 20°, but there is the opposite. This is because the temperature difference would be maximized if the slope of the collector in accordance with the zenith angle. In this study, the zenith angle of collector surface of 14.8° . On December 6, 2005 a large temperature difference at an angle 20° is greater than the angle 30°. This is because the angle of 20° approaching the zenith angle than angle 30°. On December 7, a large temperature difference at an angle of 20° mostly the result is higher than the temperature difference 40°, this is because the angle zenith angles approaching 20° more than 40°. But at the corner of the graph shows the presence of some value at an angle of 20° that the result is smaller than the angle of 40°, this is due to return air stream. It also can lead basar intake temperature becomes greater than the output temperature.

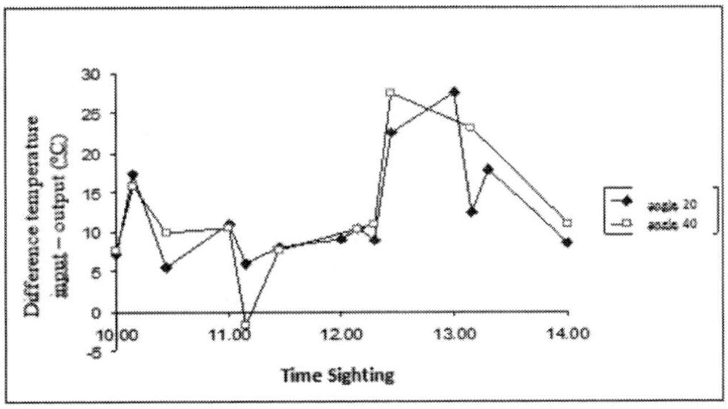

Figure 6. Chart input-output temperature difference at the corner of 10° and 20°

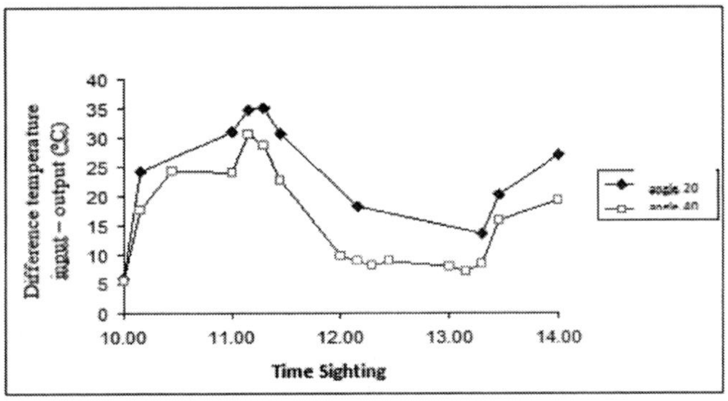

Figure 7. Chart input-output temperature difference at the corner of 20° and 30°

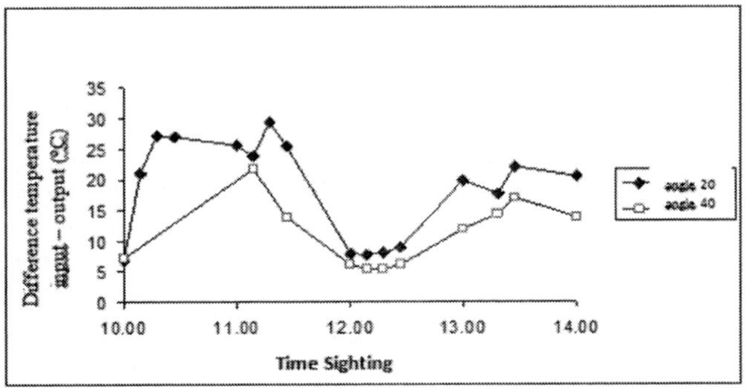

Figure 8. Chart input-output temperature difference at the corner of 20° and 40°

From the above results it can be concluded that the variation of the angle effect on the temperature difference collector. Where in the temperature difference will be maximum if the collector surface perpendicular to the sun's position [8].

3.1.3 Efficiency of solar collection on variations within the glass

The results of the efficiency of the solar collector can be seen in the graph efficiency solar collector to the temperature difference between the entry (T_{in}) and the discharge temperature (T_{out}), which is in Figure 9, Figure 10, and Figure 11.

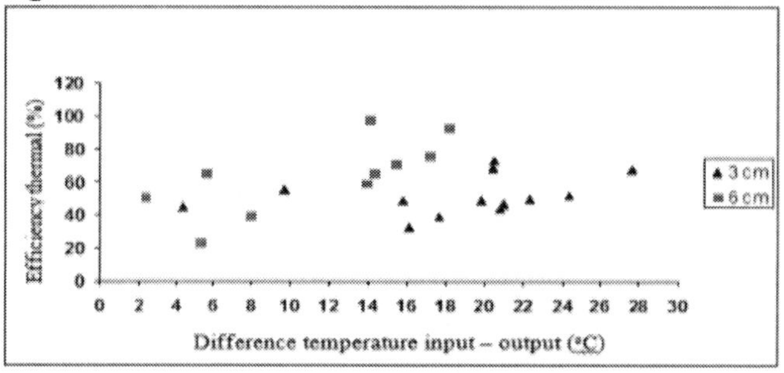

Figure 9. Graph thermal efficiency with a distance of absorbent glass cover plate with 3 cm and 6 cm

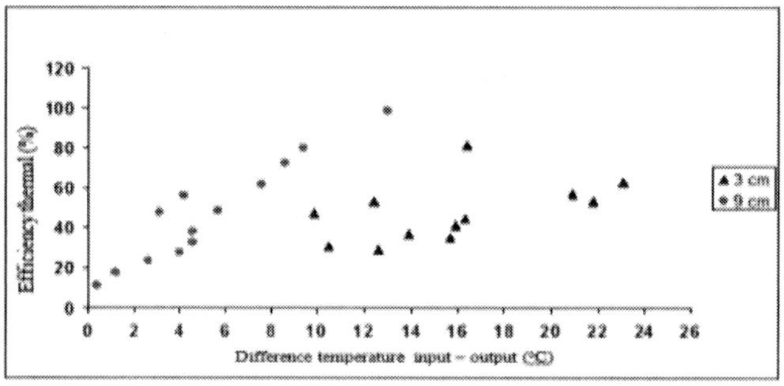

Figure 10. Graph thermal efficiency with a distance of absorbent glass cover plate with 3 cm and 9 cm

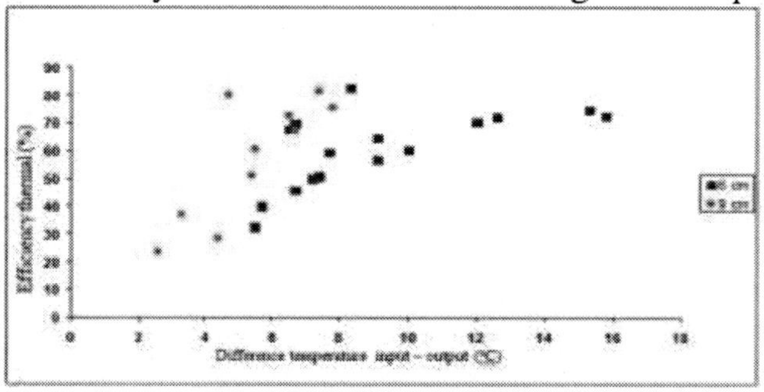

Figure 11. Graph thermal efficiency with a distance of absorbent glass cover plate with 6 cm and 9 cm

In Figure 11, Figure 12, and Figure 13 we can see that on November 30, the highest thermal efficiency of the glass at a distance of 3 cm reached 72.82% and the lowest was 33.05%. Meanwhile, at a distance of 6 cm glass highest thermal efficiency reached 97.59% and the lowest was 23.65%.

On 1 December, the highest thermal efficiency at a distance of 3 cm glass reached 81.58% and the lowest was 29.22%. Meanwhile, at a distance of 9 cm glass highest thermal efficiency reached 98.59% and the lowest was 11.2%.

On December 3, the highest thermal efficiency of the glass at a distance of 6 cm reached 82.48% and the lowest was 28.47%. Meanwhile, at a distance of 9 cm glass highest thermal efficiency reached 81.51% and the lowest was 23.6%.

So it can be said that the results of the calculation of the thermal efficiency of the solar collector in this study is not a constant but a variable characteristic that depends on the intensity of the sun, the temperature of the input, output temperature and airflow. Where the sun intensity received collector does not fluctuate, laminar air flow, and the temperature difference between the input and maximum output. The temperature difference will be maximum at a distance of a small glass.

3.1.4 Efficiency solar collector on the variation of the angle of the collector

The results of the efficiency of the solar collector can be seen in the graph efficiency solar collector to the temperature difference between the entry (T_{in}) and the discharge temperature (T_{out}), that in Figure 12, Figure 13, and Figure 14.

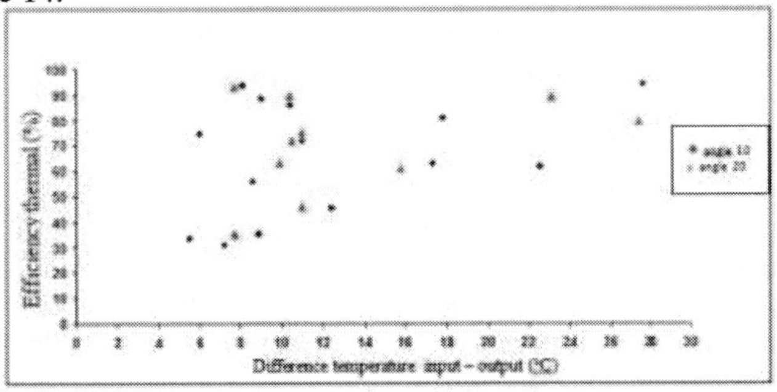

Figure 12. Graph thermal efficiency with an angle of 10° and 20°

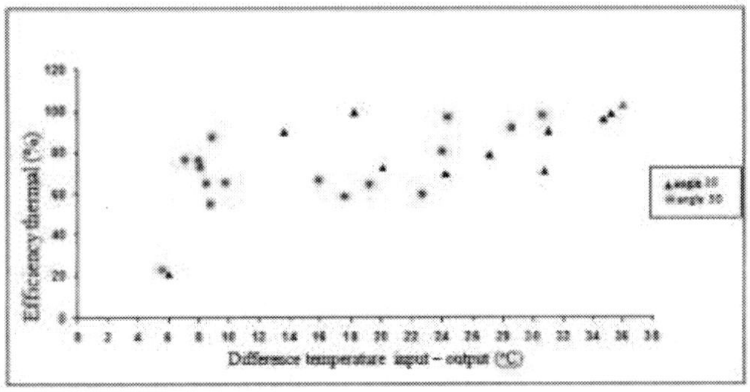

Figure 13. Graph thermal efficiency with an angle of 20° and 30°

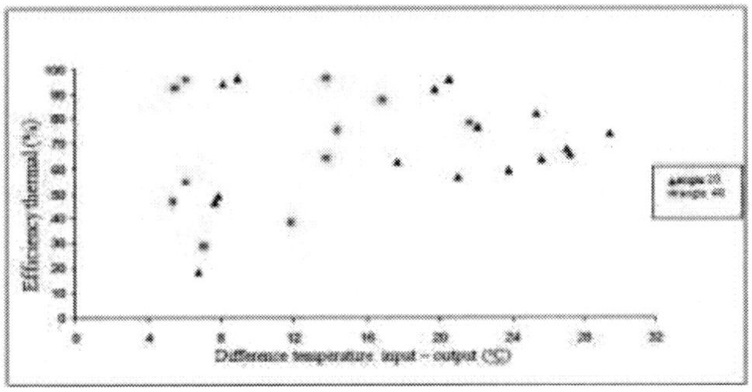

Figure 14. Graph thermal efficiency with an angle of 20° and 40°

In Figure 14, Figure 15, and Figure 16 we can see that on 5 December, the highest thermal efficiency at an angle of 100 reached 94.46% and the lowest was 31.26%. While in the corner of the 200 highest thermal efficiency reached 93.04% and the lowest was 35.23%.

On 6 December, the highest thermal efficiency at an angle of 200 reached 99.23% and the lowest was 20.92%. While in the corner of the 300 highest thermal efficiency reached 97.53% and the lowest was 22.64%.

On 7 December, the highest thermal efficiency at an angle of 200 reached 96.29% and the lowest was 18.10%. While in the corner of the 400 highest thermal efficiency reached 96.43% and the lowest was 28.5%.

So it can be said that the results of the calculation of the thermal efficiency of the solar collector in this study is not a constant but a variable characteristic that depends on the intensity of the sun, the temperature of the input, output temperature and airflow. Where the sun intensity received collector does not fluctuate and collector surface perpendicular to the sun's position, laminar air flow, and the temperature difference between the input and maximum output. The maximum temperature difference would be if the collector surface perpendicular to the sun's position.

4. Conclusion

In the third variation within absorber plate with a transparent glass, a score difference of input-output temperature at a distance of 3 cm highs and lows at a distance of 9 cm, and the absorber plate absorbs solar radiation optimally if the position of the plate perpendicular to the direction of the radiation coming sun. The slope of solar collector closer zenith angle of the input-output temperature difference increases. The thermal efficiency depends on the intensity of the sun, the temperature of the input, output temperature, thermal efficiency and airflow.

Acknowledgment

The authors would like to acknowledge The Ministry of Research, Technology and Higher Education (Kemenristek-Dikti) THE REPUBLIC OF INDONESIA for financial support in this project.

References

[1] Karyono T.H 2015 *Journal Buildings* **5** 802-813
[2] Welss W and Buchinger J 2002 *Solar Drying* AEE Intec Institute for Sustainable Technologies
[3] Bridle R., Kiston L and Wooders P 2014 *Fossil-Fuel Subsidies: A barrier to renewable energy in five Middle East and North African countries* Published by the International Institute for Sustainable Development
[4] Tharakan P 2015 *Summary of Indonesia's energy sector assessment* Published by Asian Development Bank
[5] Duttle J A and Beckam W A 1991 *Solar Engineering of Thermal Process* John Willey and Sons Inc Wisconsin
[6] Wiranto Arismunandar 1985 *Teknologi Rekayasa Surya* edisi pertama PT Pradnya Paramita Jakarta.
[7] Beiser A. 1995 *Concept of Modern Physics* 5th edition Mc Grow Hill New York.
[8] Gunther M 2002 *Solar Radiation* Advanced CSP Teaching Materials EnerMena.

Improvement of Thermo-Mechanical Properties of Short Natural Fiber Reinforced Recycled Polypropylene Composites through Double Step Grafting Process

O A Saputra[1], K S Rini [1], T D Susanti[1], R E Mustofa[1], M D Prameswari[1], E Pramono[1]

[1]Chemistry Department, Faculty of Mathematic and Natural Sciences, Sebelas Maret University Surakarta, Indonesia.

E-mail: oziadisaputra@student.uns.ac.id

Abstract. This study focused on the effect of a compatibilizer addition, maleic anhydrides (MAH) on mechanical, thermal and water absorption properties of oil palm empty fruit bunches (EFB) fiber reinforced recycled polypropylene (rPP) biocomposites. The double steps grafting process were conducted by incorporated MAH on both rPP and EFB to improve the surface adhesion between these materials, to result in a good mechanical properties as well as biocompatibility to nature. The chemical test was carried out using FTIR (Fourier Transform Infra-Red) spectroscopy technique to evaluated grafting process. The mechanical test was investigated and found that the addition of 10 phr MAH to both rPP and EFB improved mechanical strength of the biocomposites higher than another formulas. In this study, thermal properties of biocomposites also characterized. Water absorption (WA) analysis showed the presence of EFB fiber increased the water uptake of the material.

1. Introduction

Natural fiber reinforced polymer has been interested to developed due to their advantages such as excellent mechanical properties, biodegradable, renewable, lightweight, economical and environmentally friendly [1,2]. Moreover, developing of natural fiber reinforced polymer was motivated by an environmental issue which required green material for sustainability [3]. Polypropylene was hydrophobic polymer materials which mostly used in several applications such as packaging, interior automotive panels, aerospace and others [4,5]. To improve their properties, fiber reinforcing materials, such as natural fiber was introduced. However, it is difficult to obtain good adhesion between hydrophilic cellulose and hydrophobic polypropylene due to their different polarity. This problem will lead the final product has poor mechanical properties due poor ability to stress transfer from matrix polymer to fiber [6]. Additionally, cellulose fiber has highly hydrophilic which lead strong hydrogen bonding between the fiber, thus resulted in poor dispersion of fiber in the matrix and weak interfacial interaction between fiber and matrix [3,7].

The previous researchers have been modified the natural fiber surfaces to improve their adhesion between polymer and fiber. This step is one of promising strategy to obtain high strength materials. The

[1] Chemistry Department, Faculty of Mathematic and Natural Sciences, Sebelas Maret University Surakarta, Indonesia.

grafting agents have been used by previous researcher, i.e. silane compound [8], acrylic acid [9], maleic anhydrides [10, 11] and others. Jiang et al. [8] was modified the fiber surface with [3(2-aminoethyl) aminopropyl] triethoxysilane on the polyurethane/carbon fiber (CF) composites preparation and improve tensile strength up to 42% compare to composites without grafting agent. Suharty et al. [9] used acrylic acid as grating agent on the recycled-polypropylene (rPP)/rice husk powder (RHP) biocomposites to obtain high mechanical properties. Acrylic acid grafted polypropylene (PP-g-AA) was also introduced by Suharty et al. [12] on the manufactured rPP/kenaf fiber (KF) biocomposites, resulted high mechanical properties. Arao et al. [13] added maleic anhydrides grafted polypropylene (PP-g-MAH) on PP/CF composites, obtained an improvement of interfacial share strength up to 119% compared to PP. The PP-g-MAH grafting agent has polar maleic group which can forming ester linkage with hydroxyl group of natural fiber. In addition, polypropylene backbone of PP-g-MAH will be form one phase with polypropylene chain [3].

In our previous research [10], has been successfully manufactured rPP/short empty fruit bunch fiber (sEFB) with the presence of PP-g-MAH. However, the MAH grafting process was only performed on polypropylene chain. Thus, in this research was conducted two steps grafting process which are both in rPP and sEFB. This research was focused on the effect of two steps grafting process on thermo-mechanical properties of rPP/sEFB biocomposites.

2. Experimental

2.1. Materials

Recycled polypropylene (rPP) from waste plastic bottle water which collected from Surakarta, Indonesia was used as matrix polymer. Short natural fibers was oil palm empty fruit bunches (EFB) which obtain from LIPI Serpong, Indonesia. Compatibilizer maleic anhydride (MAH) was obtained from Merck. Other chemicals such as benzoyl peroxide (BPO, Merck) and divinyl benzene (DVB, Merck), xylene (PT. Brataco, Indonesia), toluene (Merck), acetone (Merck), ethanol (Merck), N,N-dimethyl formamide (DMF, Merck), hydroquinone (purchase from UMS Surakarta) were used without purification.

2.2. Preparation of MAH grafted EFB (EFB-g-MAH)

The grafting of EFB fiber was prepared according to Rozman et al. [14] method. The EFB fiber was chopped about 2-5 mm and immerse in a mixture solution of toluene, ethanol and acetone with ratio 4:1:1 about 3 hours. The fibers were filtered and dry about 16 hours in temperature 105 °C. In the different place was prepared variation concentration of maleic anhydride (5, 10, 15 and 20 phr w/w) by dissolved it in DMF. Afterward, hydroquinone (5% w/w to MAH) was added to MAH solution. The solution was added drop-wise into three round neck bottom flask contained 50 grams of EFB. The mixture was refluxed about 1 hour at 80 °C. Grafting product t-EFB-g-MAH was washed using acetone to remove unreacted MAH. The product was dried about 16 hours at 105 °C to obtain EFB-g-MAH. The raw material and the final product was characterized using FTIR.

2.3. Fabrication and testing of rPP/EFB biocomposites

Table 1. showed a formulation in the manufacturing of rPP/sEFB bio-composites. Fabrication of rPP/EFB biocomposites was conducted using solution method and radically reaction. About 20 grams of rPP was dissolved in boiling xylene. The previous optimum amount of MAH (10% w/w), benzoyl peroxide and divinyl benzene were added after rPP successfully melted. After 5 minutes, the EFB-g-MAH was added and stirred for 1 hours. The rPP/sEFB bio-composites was dried for 24 hours at room temperature and fabricated using hot press hydraulic machine.

Table 1. The formulation of rPP/EFB biocomposites.

Materials	Formulas (phr)						
	F1a	F1b	F2a	F2b	F2c	F2d	F2e
rPP	100	100	70	70	70	70	70
EFB	-	-	30	-	-	-	-
EFB-g-MAH5	-	-	-	30	-	-	-
EFB-g-MAH10	-	-	-	-	30	-	-
EFB-g-MAH15	-	-	-	-	-	30	-
EFB-g-MAH20	-	-	-	-	-	-	30
MAH	-	10	-	10	10	10	10

Thermal properties was analyzed using Simultaneous Thermal Analyzer (STA) Lineises PT-1600. The sample was recorded in the range of 30-600 °C with the heating rate of 10 °C/minute. Before analysis, the sample was dried in an oven at 50 °C for 1 hour.

Mechanical properties of biocomposites were evaluated by following ASTM D638-02a method. The tensile test was performed using mechanical testing machine Ray-Ran M500-50CT with crosshead speed of 10 mm/minute.

Water adsorption was carried out by measure the mass of sample after immersed in water for 1 hour. The samples before testing were weighed and recorded as initial weight (Wo). After immersion for 24, 48, 72, 96 and 120 hours, the samples were weighed again and recorded as final weight (Wi), then determined the percentage of water absorption capacity (% WA) by equation 1.

$$\%WA = \frac{(Wi-Wo)}{Wo} \times 100\% \quad \ldots\ldots(1)$$

3. Result and Discussion

3.1. Functional group analysis

The grafting product of EFB fiber was characterized using FTIR technique. Fig 1 showed FTIR spectra of EFB and EFB-g-MAH. A strong absorbance band at 1743 cm^{-1} in EFB-g-MAH spectrum was corresponding to the carbonyl group of MAH. The appearance of this absorbance band indicated that grafting process was successfully achieved. The low peak intensity of hydroxyl group after after grafting was affect by ester linkage formation. The hydroxyl group of EFB was react with anhydride group of MAH to form an ester linkage.

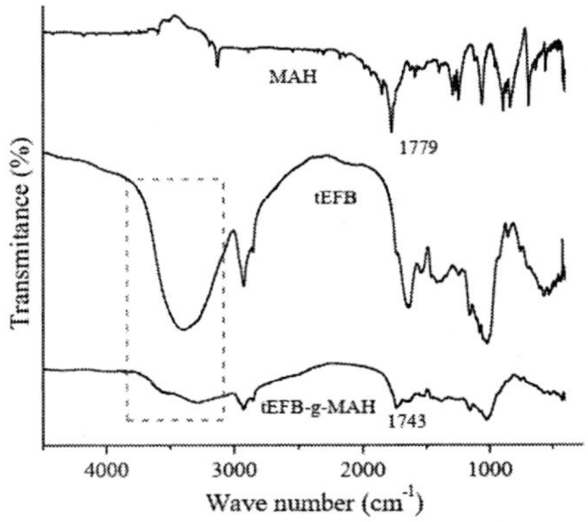

Figure 1. FTIR spectrum of MAH, EFB and EFB-g-MAH

3.2. Mechanical properties of the biocomposites

Mechanical properties of some composites with and without compatibilizer maleic anhydride had been showed in Fig 2. It was noted from our previous work that the uncompatibilizing rPP/tEFB (F2a) bio-composites have lower tensile strength compared to others formula, 24.67 MPa. Surface treatment of EFB fiber by the addition of MAH was able to improve the mechanical properties. By using of 10% MAH into EFB fiber was able to increase tensile strength up to 84 % compared to ungrafted EFB fiber. The presence of MAH in EFB fiber improved their surface adhesion with hydrophobic polypropylene matrix. Good adhesion between filler and matrix caused that optimum stress transfer from matrix to fiber was achieved. Consequently, the tensile strength value of the biocomposites will be increased. However, the tensile strength by the addition of 15 and 20 phr of MAH was slightly decreased. It indicated that surface modification of EFB with maleic anhydride has optimum state at 10 phr MAH addition.

Elongation data is one of indicators to determine the elasticity of materials. The higher value of materials elongation indicated that it has high elasticity. From the data, by introduced MAH in composites system was not only improve their strength, but also their elasticity. The F2a composites has elongation under 20%, however, almost composites contained MAH had elongation above 20%.

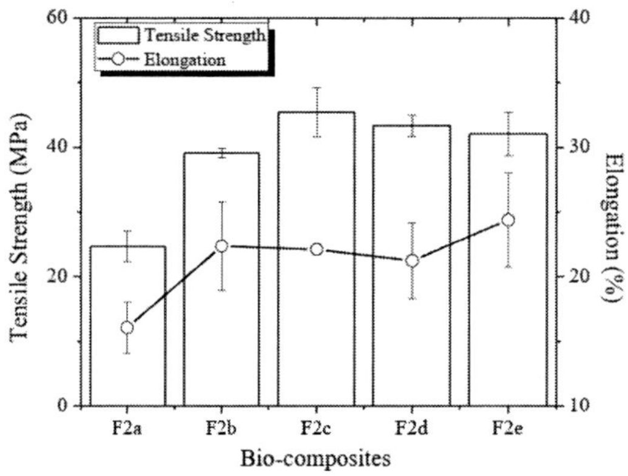

Figure 2. Tensile strength and elongation of the biocomposites

3.3. Thermal properties

Thermal properties of untreated and grafted fiber on rPP/tEFB biocomposites has been shown in Fig 3. Wight loss of rPP/tEFB bi-composite in the first degradation was about 74%. However, after surface grafting was applied in the tEFB fiber, the weight loss of bio-composite was reduced to about 59%. This indicated that by surface treatment on EFB fiber improved the thermal stability of biocomposites. Good interaction between fiber and matrix will improve the bond strength of the composites, thus it was impact on the improving of thermal stability. Different thermal analysis (DTA) thermograph showed that the second degradation temperature of biocomposite contained EFB-g-MAH10 slightly increased compared to untreated fiber, which was 445 °C. The possible reason for this phenomena is the fiber has hydrophilic properties will lead the fiber has high hydrogen bonding formation. The internal hydrogen bonding between hydroxyl groups in tEFB fiber with MAH groups was higher than without MAH groups. Thus, it required high energy to degrade the fiber with the presence of MAH.

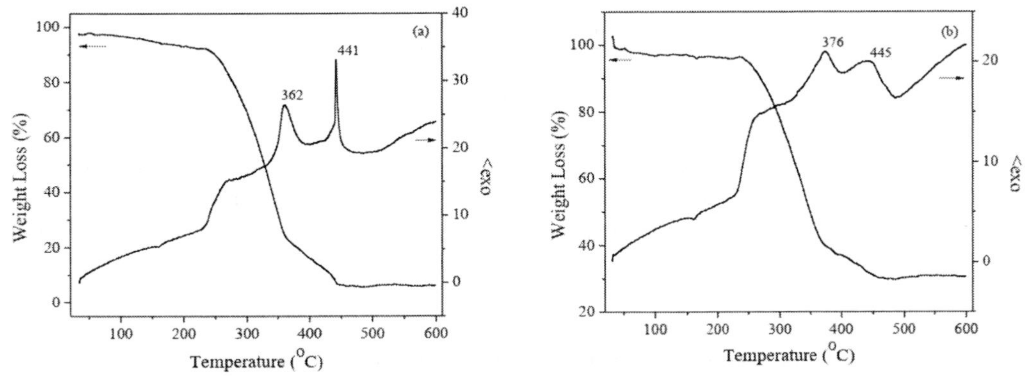

Figure 3. TGA graph of (a) rPP/EFB and (b) rPP/EFB-g-MAH10 biocomposites

3.4. Water absorption properties

Water adsorption behavior of unmodified rPP/EFB and rPP/tEFB-g-MAH10 biocomposites was showed in Fig 4. The rPP/EFB-g-MAH10 biocomposites displayed an increasing of water adsorption by soaking for 48 hours. Afterward, the composite reached saturation adsorption at immersion time of 72 and 96 hours, about 8.5%. It was occurr due to their limited water uptake ability which causing by good distribution and interaction between matrix and the fiber and resulting in the improvement of interfacial adhesion with the presence of compatibilizer agent [15]. This good interaction resulted in low void content in the biocomposites, while the void content was one of water adsorption factors. Also, good interaction between matrix and the fiber will lead the fiber difficult to interact with water. By comparing to unmodified rPP/tEFB bio-composites, the water adsorption value increased with the increasing of immersion times, while after immersing at 24, 48, 72 and 96 hours, the biocomposites was allowed to adsorb the water up to 4.3%, 13.1%, 21.7% and 23.9%, respectively. In rPP/EFB biocomposites, saturation adsorption was not actually reached even after 96 hours immersion times. Water adsorption was influenced by several factors such as fiber content, void content, fiber orientation, surface protection and many others [3, 16].

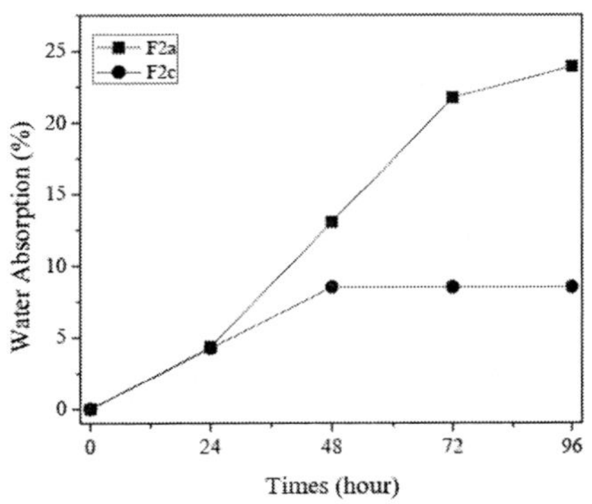

Figure 4. Water adsorption of rPP/EFB and rPP/EFB-g-MAH10

Cellulose fiber has hydrophilic in nature, it means that, they was easy to absorb the water due to the hydrogen bonding between an organic compound and water molecules. Without compatibilizer agent, the matrix and the fiber has poor interaction and lead to produce the void site. Consequently, it was able to entrap the water molecules [17]. Moreover, high polar fiber fraction was easy to absorb the water and increase the composites weight. It can be concluded that water adsorption behavior of some biocomposites materials was dependent on the presence of free hydroxyl group on the surface of fiber

[18]. Thus, by the addition of MAH into the fiber, esterification between anhydride group and hydroxyl group occurred. As the result, the free hydroxyl group on the fiber surface decreased.

4. Conclusion
The grafting of both short natural fiber and polypropylene on the rPP/EFB composites have significantly influence on both mechanical and thermal properties. Surface grafting of short natural fiber with 10 phr maleic anhydride is the optimum composition on the composites preparation due to they has high mechanical properties compared to another formula. In another cases, the presence of MAH also decreased the water adsorption of the composites.

Acknowledgement
Authors would like to acknowledge to Indonesian Directorate General of Higher Education for financial support from Program Kreatifitas Mahasiswa (PKM) 2014. We also gratefully acknowledge to Diknas Jawa Tengah to support this project through PPIM programs.

References
[1] Arbelaiz A, Fernandez B, Ramos J A, Retegi A, Llano-Ponte R, Mondragon I 2005 *Compos. Sci. Technol.* **65** 1582
[2] Faruk O, Bledzki A K, Fink H P, Sain M 2012 *Progg. Polym. Sci.* **37** 1552
[3] Osman H, Ismail H, Mustapha M 2010 *J. Compos. Mater.* **44** 1477
[4] Rezaei F, Yunus R, Ibrahim N A, Mahdi E S 2008 *Polym. Plas. Technol. Eng.* **47** 351
[5] Ayrilmis N, Jarusombuti S, Fueangvivat V, Bauchongkol P, White R H 2011 *Fiber Polym.* **12** 919
[6] Karmarkar A, Chauhan S S, Modak J M, Chanda M 2007 *Compos. Part A: Appl. Sci. Manuf.* **38** 227
[7] Sombatsompop N, Yotinwattanakumtorn C, Thongpin C 2005 *J. Appl. Polym. Sci.* **97** 475
[8] Jiang S, Li Q, Zhao Y, Wang J, Kang M 2015 *Compos. Sci. Technol.* **110** 87
[9] Suharty N S, Mathialagan M, Ismail H, Wirjosentono B, Firdaus M, Wardani G K 2014 *J. Physic. Sci.* **25** 55
[10] Saputra O A, Rinawati L, Rini K S, Saputra D A, Pramono E 2016 *Appl. Mechan. Mater.* **842** 7
[11] Li M, Wen X, Liu J, Tang T 2014 *Compos. Part A: Appl. Sci. Manuf.* **67** 212
[12] Suharty N S, Ismail H, Handayani D S, Diharjo K, Wibowo F R, Wuri M A 2016 *IOP Conf. Series: Mater. Sci. Eng.* **107** 012038
[13] Arao Y, Yumitori S, Suzuki H, Tanaka T, Tanaka K, Katayama T 2013 *Compos. Part A: Appl. Sci. Manuf.* **55** 19
[14] Rozman H D, Saad M J, Mohd Ishak Z A 2003 *J. Appl. Polym. Sci.* **87** 827
[15] Thwe M M, Liao K 2002 *Compos. Part A: Appl. Sci. Manuf.* **33** 43
[16] Ashori A, Nourbakhsh A 2008 *Polym. Compos.* **5** 574
[17] Nachtigall S M B, Cerveira G S, Rosa S M L 2007 *Polym. Test.* **26** 619
[18] Osman H, Ismail H, Mariatti M 2012 *Polym. Compos.* **33** 609

IOP Conf. Series: Earth and Environmental Science **75** (2017) 012024 doi:10.1088/1755-1315/75/1/012024

Surface Properties and Permeability of Poly(Vinylidene Fluoride)-Clays (PVDF/Clays) Composite Membranes

E Pramono[1,3], M Ahdiat[1], A Simamora[1], W Pratiwi[1], C L Radiman[1] and D Wahyuningrum[2]

[1]Inorganic and Physical Chemistry Division, Faculty of Mathematics and Natural Sciences, Institut Teknologi Bandung, Jalan Ganesha 10, Bandung, Indonesia

[2]Organic Chemistry Division, Faculty of Mathematics and Natural Sciences, Institut Teknologi Bandung, Jalan Ganesha 10, Bandung, Indonesia

[3]Inorganic Materials Research Group, Faculty of Mathematics and Natural Sciences, Sebelas Maret University, Jalan Ir Sutami 36A, Surakarta, Indonesia

E-mail: cynthia@chem.itb.ac.id

Abstract. Surface properties are important factors that determine the performance of ultrafiltration membranes. This study aimed to investigate the effects of clay addition on the surface properties and membrane permeability of PVDF (poly-vinylidene fluoride) membranes. Three types of clay with different particle size were used in this study, namely montmorillonite-MMT, bentonite-BNT and cloisite 15A-CLS. The PVDF-clay composite membranes were prepared by phase inversion method using PEG as additive. The hydrophobicity of membrane surface was characterized by contact angle. The membrane permeability was determined by dead- end ultrafiltration with a trans-membrane pressure of 2 bars. In contact angle measurement, water contact angle of composite membranes is higher than PVDF membrane. The addition of clays decreased water flux but increased of Dextran rejection. The PVDF-BNT composite membranes reach highest Dextran rejection value of about 93%. The type and particle size of clay affected the hydrophobicity of membrane surface and determined the resulting membrane structure as well as the membrane performance.

1. Introduction

PVDF membranes have been widely applied in water treatment processes. PVDF membranes have many advantages including high mechanical properties and high chemical resistance [1, 2]. During their applications, low resistance to fouling often limits the use of these membranes [3]. Fouling on the membrane surface decreases its water permeability due to its surface properties.

Several factors affect the performance of an ultrafiltration membrane such as surface structure, pore size and membrane structure [4, 5]. Surface properties like membrane polarity affect the ease of interaction of water or compounds that are separated in the feed solution [6-8]. Some researchers have developed various surface modifications of PVDF membranes to produce membranes with high performance. Chemical reaction with a plasma method and addition of zwitter ionic compound which modified the surface and pore properties and its performance against the Bovine Serum Albumin (BSA) had been reported [9, 10]. Unfortunately, those modifications are often limited by hard reaction and low interference between the components of the membrane.

Fabrication of composite PVDF with inorganic filler was also developed by many researchers [11-16]. This method is widely conducted because it provides many advantages such as easy preparation, ability to modify the surface properties and improvement of membrane mechanical and thermal properties [17-19]. Some oxides had been used as fillers in PVDF membranes. Previous study showed that the addition of Mg(OH)2 in PVDF membranes can improve the fouling resistance against BSA and E. coli. [20]. Modification of PVDF with palygorskite could increase the water flux and also the abrasion of membrane surface [21].

In this study various clays were used as filler in PVDF membranes. The effects of types of clay on the surface properties and permeability of PVDF membranes have been studied. Three types of clays having different particle size were used in this research, which are bentonite (BNT), montmorillonite (MMT), and cloisite (CLS) 15A. Membranes were prepared by phase inversion method and surface properties of membranes were characterized by contact angle and surface morphology using a Scanning Electron Microscope (SEM). The membrane permeability and selectivity were characterized by water flux and rejection towards Dextran T-500.

2. Experiments

2.1. Materials
PVDF powder (Solef® 1015, solvay), polyethylene glycol (PEG)-400 (Merck), ethylene glycol (99%, Merck), n-hexane (99%, Merck), dimethylacetamide – DMAc (99%, Merck) and Dextran T-500 (Mw 500 kDa) (Pharmacosmos) were used without further purification. Three types of clay were used, which are bentonite (BNT) 180 mesh supplied by a local industry, montmorillonite and cloisite 15 A (Sigma Aldrich) with a particle size of 8.59; 13 and 24.11 μm, respectively.

2.2. Membrane Preparation
PVDF and PVDF-clay composite membranes were prepared by phase inversion method. Measured amount of PVDF, PEG 400 and clay were mixed in DMAc solvent with a composition of 17%, 5% and 1% w/w, respectively. The mixture was stirred at 60 °C for 5 h until homogeneous. The polymer mixture was sonicated with frequency 47 kHz for 1 h. The solution was spread on a glass plate and directly immersed into a coagulation bath containing water. The solid flat membranes were washed under running water to remove the residual solvent. The membrane was then stored in glycerine for further characterization. The symbols of PVDF-BNT, PVDF-CLS and PVDF-MMT denoted the composite membranes using the corresponding type of clay.

2.3. Membrane morphology and measurement of contact angle
The morphology of the surface and cross section of membranes were observed by a scanning electron microscope (JEOL JSM-6360LA) with a magnification of 5000 and 1000, respectively. The surface hydrophobicity of the membranes was characterized by contact angle (θ) measurement using drop analysis method. A piece of flat membranes was put on the sample holder. One drop of deionized water was dropped and the image was recorded. The contact angle of membrane was also measured using ethylene glycol and n-hexane as the liquid. All measurements were carried out at room temperature and the resulting images were analyzed with Image J software using drop analysis method.

2.4. Membrane characterization
The pure water flux and rejection coefficient of membranes were determined using a dead-end ultrafiltration equipment. Prior to measurement, membranes with 5 cm diameter were washed with pure water to remove glycerin. The membrane was compacted at 2 bars to get a steady flux and the permeate was collected every 5 mins; at least three data was taken to obtain an average flux for each type of membrane. The water flux was calculated using the following equation:

$$J = \frac{V}{A \times t} \quad (1)$$

where J is the water flux of membrane (L.m^{-2}.h^{-1}), V is the volume of the permeate (L), A is the area of the membrane (m^2), t is the permeation time (h).

The rejection percentage was measured using the same equipment as for water flux measurement, but the feed was a T-500 Dextran solution with a concentration of 1000 ppm. The rejection was calculated as follows:

$$R = \left(1 - \frac{C_P}{C_R}\right) \times 100 \quad (2)$$

where R is the rejection of the membrane (%), C_p and C_R are the concentration of permeate and feed solution, respectively.

3. Results and discussion

3.1. Membrane wettability

The membrane wettability can be examined by contact angle using polar and non-polar liquids. In this research, the wettability of PVDF and PVDF-clay composite membranes were analysed using polar liquids, which are water and ethylene glycol, and non-polar liquid such as n-hexane. The contact angles of water on the surface of various composite membranes were shown in Figure 1. It can be seen that the contact angles of PVDF-BNT and PVDF-CLS composite membrane were higher than the one of the pristine membrane. On the other hand, the contact angle of PVDF-MMT membrane was the lowest. The addition of BNT and CLS reached contact angle 69.2° and 76.7° respectively. It should be noted that all types of clay used in this study have similar main structure as MMT which contain negative charge in their structure. Natural BNT is MMT which still contains other natural oxides, while cloisite is MMT which has been intercalated by the addition of quaternary ammonium salt. The presence of clay on membrane surface affects the interaction between membrane and water, and also the surface roughness of membrane. The addition of oxide as a membrane component can increase the roughness of the membrane, and this has an impact on increasing the membrane hydrophobicity. On the other hand, the addition of MMT decreased the contact angle of water; it may be due to the polar nature of MMT.

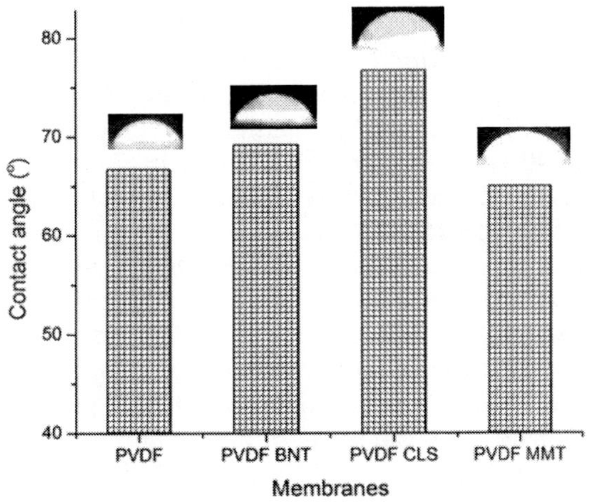

Figure 1. Contact angle of water on the surface of various composite membranes

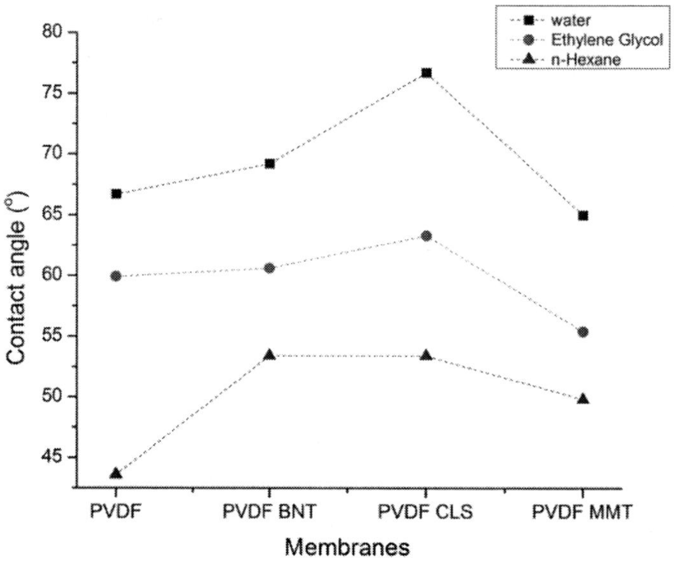

Figure 2. Membranes wettability of various composite membranes towards polar and non-polar liquids

Figure 2 showed membrane wettability towards polar and non-polar liquids. The contact angles of ethylene glycol have the same pattern with the one of water, while n-hexane showed a different trend. All the composite membranes gave higher values of contact angle than the pristine PVDF membrane when n-hexane was used. N-hexane contact angle of PVDF-BNT, PVDF-CLS and PVDF-MMT increased from 43.6° to 53.4 °, 53.4 ° and 49.8 ° respectively. Since PVDF membrane is non-polar, its interaction with n-hexane is high, resulting in a relatively small contact angle towards n-hexane. It can be said that the addition of clay changed the surface structure of membrane and the polar properties of the membranes as well.

3.2. Membrane permeability and selectivity

Membrane permeability and selectivity were determined indirectly by measuring the water flux and rejection towards Dextran T-500. Effect of the addition of clay towards the water flux and the rejection toward Dextran T-500 solution were represented in Figure 3. The data showed that the addition of BNT and CLS in PVDF increased the water flux. PVDF-BNT and PVDF-CLS reached value peak of 13.05 $L.m^{-2}.h^{-2}$ and 18.15 $L.m^{-2}.h^{-2}$ respectively, and increased 65% and 125% from PVDF membrane. The addition of oxide into the membrane tend to increase the porosity of membranes which could enhance water permeability [22]. Meanwhile, Figure 3 also showed that similar rejection towards Dextran T-500 was obtained by PVDF and composite membranes, except the PVDF-BNT membrane. PVDF-BNT membrane reached rejection value of about 93%. The higher rejection coefficient might be due to the small particle size of BNT resulted in denser membranes than the others. Particle size of inorganic filler can effect of the membrane structure [23]. Consequently, the passage of Dextran compound across the membrane was inhibited and the rejection increased.

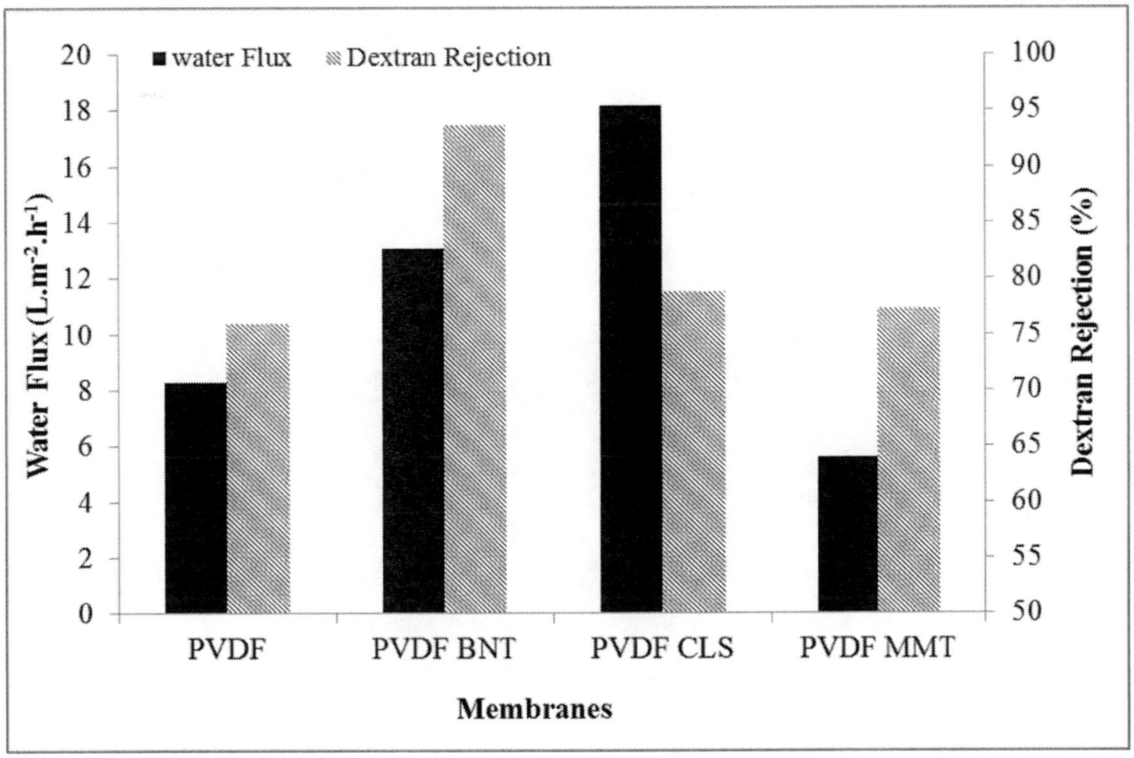

Figure 3. Water flux and Dextran T-500 rejection of pristine PVDF and various composite membranes

3.3. Membrane Morphology

The cross-section of PVDF-BNT composite membrane shown in Figure 4 indicated the presence of an asymmetrical structure; it consisted of microporous layer at the upper skin and finger-like macrovoid at the lower part of the membrane which was similar to the study by Lai *et al.* [24]. The structure is formed due to the rapid diffusion process of solvent and non-solvent during the phase inversion [25]. Figure 5 showed the difference among the morphologies of various types of membranes. The images of membrane surface looked similar one to another, but the cross-sections differed significantly. In the case of PVDF-MMT composite membrane, the addition of clay produced longer microporous layer than the pristine one which resulted in lower permeability or lower flux.

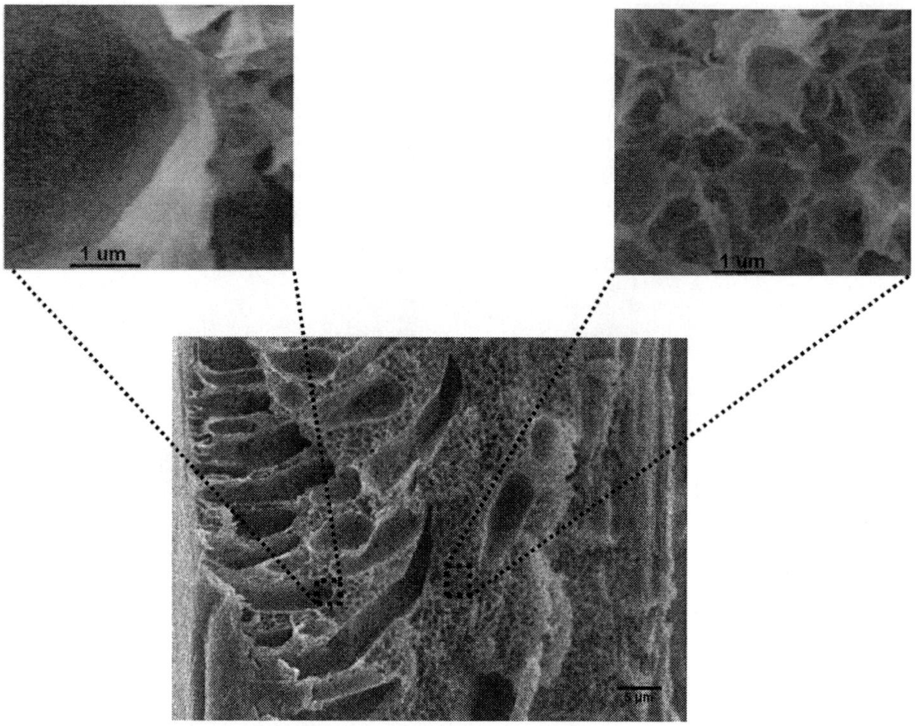

Figure 4. Cross section image of PVDF- BNT composite membrane

Membrane	Surface	Cross Section
PVDF Membrane	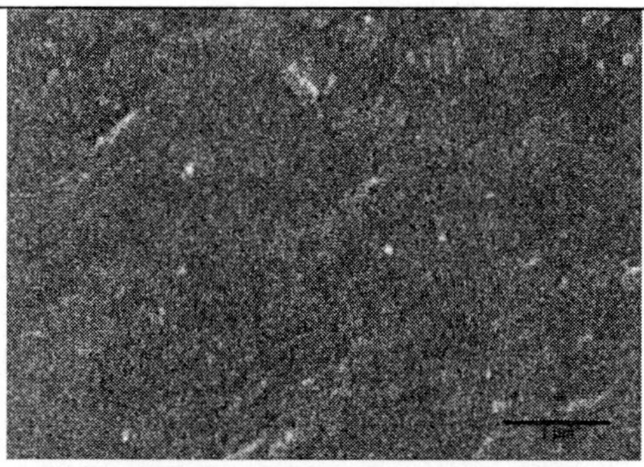	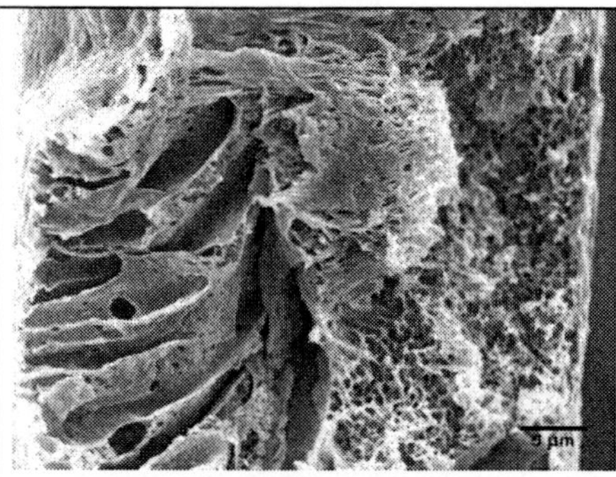
PVDF-BNT	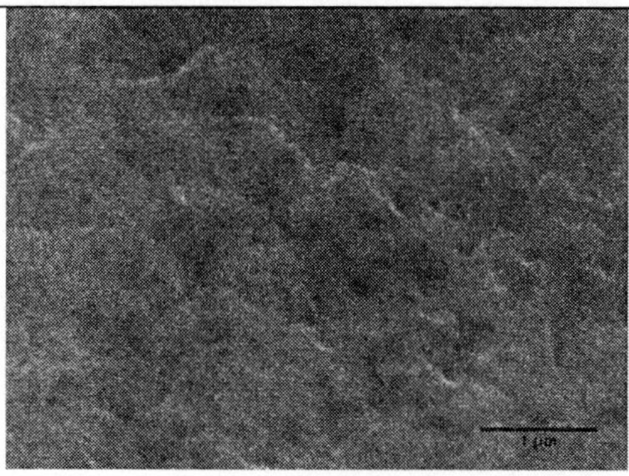	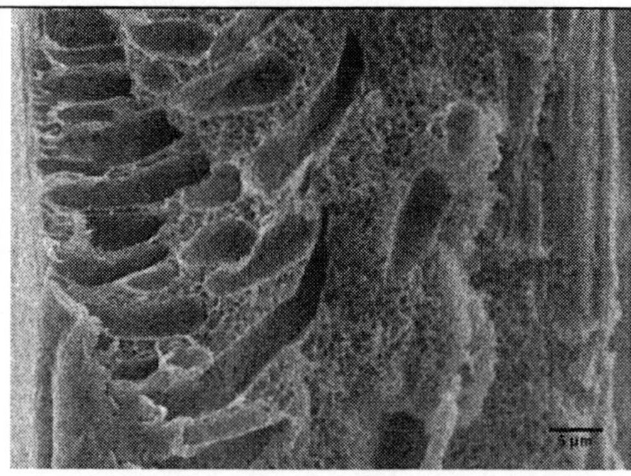

PVDF-MMT	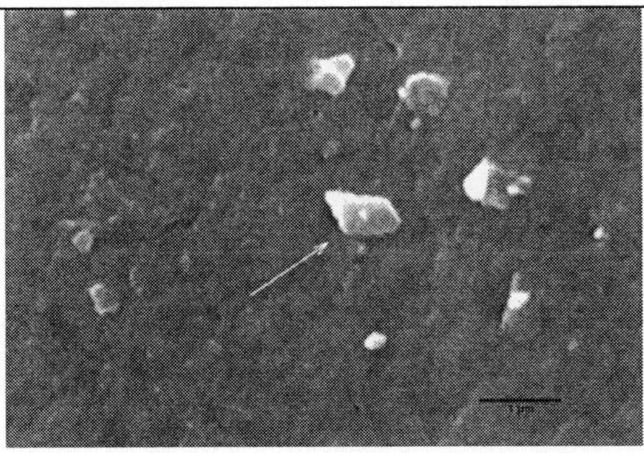	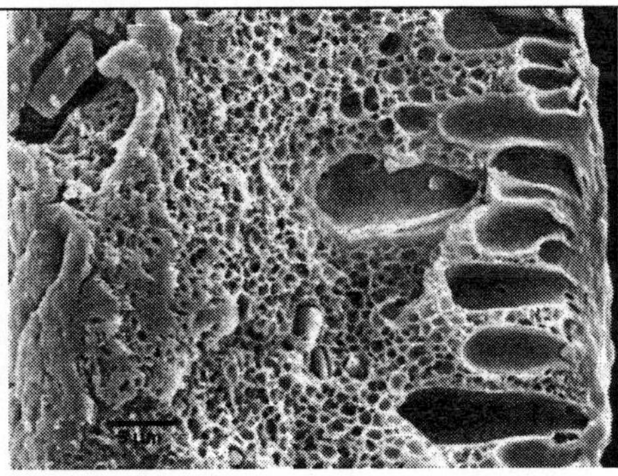

Figure 5. Surface and cross-section images of PVDF and various composite membranes

4. Conclusions

Flat sheets of various PVDF-clay composite membranes were prepared by phase inversion. The effects of clay addition on the surface properties, morphology, flux and rejection of the resulting membranes were evaluated. All membranes were found to have different asymmetric structures with spongy and finger-like pores. The addition of BNT and CLS increased the water contact angle while MMT decreased it. The negative charges of clays gave polar properties to the membrane surface, but the non-polar properties of PVDF were still dominant and resulted in lower contact angle toward n-hexane. It can be concluded that the PVDF-BNT membrane obtained the highest rejection toward Dextran T-500 (93%) compared to other composite membranes studied in this work.

Acknowledgment

The authors give the gratitude to Lembaga Penelitian dan Pengabdian Masyarakat Institut Teknologi Bandung for providing the financial support for this research through the KK Research Grant, Program A, 2016.

References

[1] Santos, W. N., Iguchi, C. Y. and Gregorio, R. 2008 *Polym. Test.* **27** 204-8.
[2] He, T., Zhou, W., Bahi, A., Yang, H. and Ko, F. 2014 *Chem. Eng. J.* **252** 327-36.
[3] Park, S. Y., Choi, S. H., Chung, J. W. and Kwak, S.-Y. 2015 *J. Membr. Sci.* **480** 122-8.
[4] Jayalakshmi, A., Rajesh, S. and Mohan, D. 2012 *Appl. Surf. Sci.* **258** 9770-81.
[5] Kim, J. H. and Lee, K. H. 1998 *J. Membr. Sci.* **138** 153-63.
[6] Kotte, M. R., Cho, M. and Diallo, M. S. 2014 *J. Membr. Sci.* **450** 93-102.
[7] Kotte, M. R., Hwang, T., Han, J.-I. and Diallo, M. S. 2015 *J. Membr. Sci.* **474** 277-87.
[8] Shen, X., Yin, X., Zhao, Y. and Chen, L. 2015 *Colloid. Poly. Sci.* **4** 1205–13.
[9] Liang, S., Qi, G., Xiao, K., Sun, J., Giannelis, E. P., Huang, X. and Elimelech, M. 2014 *J. Membr. Sci.* **463** 94-101.
[10] Zhao, X. and He, C. 2015 *Appl. Mater. Interfaces.* **7** 17947-53.
[11] Yan, L., Shui, Y., Bao, C. and Xianda, S. 2006 *J. Membr. Sci.* **276** 162-7.
[12] Baghbanzadeh, M., Rana, D., Matsuura, T. and Lan, C. Q. 2015 *Desalination* **369** 75-84.
[13] Liang, S., Xiao, K., Mo, Y. and Huang, X. 2012 *J. Membr. Sci.* **394-395** 184-92.
[14] Alpatova, A., Meshref, M., McPhedran, K. N. and Gamal El-Din, M. 2015 *J. Membr. Sci.* **490** 227-35.
[15] Wei, Y., Chu, H.-q., Dong, B.-z., Li, X., Xia, S.-j. and Qiang, Z.-m. 2011 *Desalination* **272** 90-7.
[16] Bottino, A., Capannelli, G. and Comite, A. 2002 *Desalination* **146** 35-40.
[17] Causin, V., Carraro, M. L., Marega, C., Saini, R., Campestrini, S. and Chimiche, S. 2008 *J. Appl. Polym. Sci.* **109** 2354-61.

[18] Zhang, Y. Y., Jiang, S. L., Yu, Y., Xiong, G., Zhang, Q. F. and Guang, G. Z. 2012 *J. Appl. Polym. Sci.* **123** 2595-600.
[19] Wang, P., Ma, J., Wang, Z., Shi, F. and Liu, Q. 2012 *Langmuir* **28** 4776 - 4786.
[20] Dong, C., He, G., Li, H., Zhao, R., Han, Y. and Deng, Y. 2012 *J. Membr. Sci.* **387 - 388** 40-7.
[21] Ji, J., Zhou, S., Lai, C. Y., Wang, B. and Li, K. 2015 *J. Membr. Sci.* **495** 91-100.
[22] Yang, Y., Wang, P. and Zheng, Q. 2006 *J. Polym. Sci. Polym. Phys.* **44** 879-887.
[23] Cao, X., Ma, J., Shi, X. and Ren, Z. 2006 *Appl. Surf. Sci.* **253** 2003-10.
[24] Lai, C. Y., Groth, A., Gray, S. and Duke, M. 2015 *Chem. Eng. J.* **267** 73-85.
[25] Smolders, C. A., Reuvers, A. J., Boom, R. M. and Wienk, I. M. 1992 *J. Membr. Sci.* **73** 259-75.

IOP Conf. Series: Earth and Environmental Science **75** (2017) 012025 doi:10.1088/1755-1315/75/1/012025

Diaquachloro-tris(ethyl-p-Aminobenzoate)copper(II) Chloride: Synthesis, Characterization and *In Vitro* Investigation of Antibacterial Activity

S B Rahardjo[1], W W Lestari[1], and H Syaima[1]
[1]Department of Chemistry, Faculty of Mathematics and Natural Sciences, Sebelas Maret University, Jl. Ir. Sutami 36 A Surakarta, 57126, Indonesia

E-mail: sentotbr@yahoo.com

Abstract. The new complex of $[Cu(benz)_3(H_2O)_2Cl]Cl$ has been synthesized in 1:4 mole ratio of $CuCl_2{\cdot}2H_2O$ and ethyl-p-aminobenzoate (benz) in ethanol at room temperature. The complex was characterized by UV-Vis and infrared spectroscopy, Atomic Absorption Spectroscopy (AAS), thermal analysis, magnetic measurement and molar conductivity. Infrared spectra indicated that benz was coordinated to the metal ion through nitrogen of primary amine group. Water molecules and chloride ion were also coordinated to Cu(II). $CuCl_2{\cdot}2H_2O$, benz, and Cu(II) complex were screened for investigating *in vitro* antibacterial activity against *Staphylococcus aureus* and *Escherichia coli* using a modified Kirby-Bauer method.

1. Introduction

Ethyl-p-aminobenzoate (benz), shown in figure 1, is an anesthetic that can be used to treat pain [1]. As a remedy, benz has weaknesses regarding their low solubility in water but has the ability to inhibit bacterial growth of gram-negative and gram-positive bacteria that can be developed as an antibacterial [2]. In general, the antibacterial properties of the ligands are increased after complexing [3],[4]. Functional groups of a primary amine, a carbonyl group, an oxygen atom in the C=O group and electron cloud of benzene in benz that have lone pair electrons are able to form a complex compound with a metal such us copper. Cu(II) of $CuCl_2{\cdot}2H_2O$ shows antibacterial activity which is able to inhibit the growth of *Escherichia coli* [5] due to the ability to kill bacteria called the oligodynamic action. Therefore, ligand complexed with Cu(II) has better antibacterial activity compared to ligand without complexing [6].

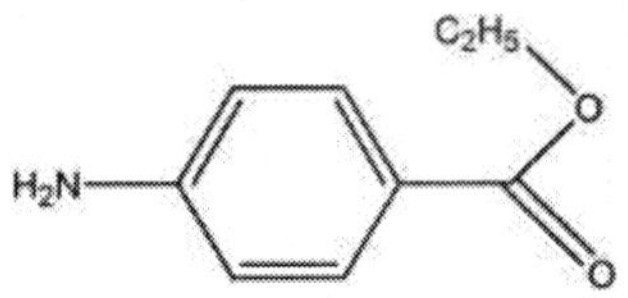

Figure 1. The structure of benz.

Cu(II) complex can form an assortment of geometries in accordance with a coordinated ligand. 8-hidroksiquinolin (HQ) and alanine (ALA) coordinated with Cu(II) form a complex [Cu(HQ)(ALA)] with tetrahedral geometries [7]. Cu(II) is coordinated to HQ through the pyridine nitrogen and oxygen from hydroxyl groups, while ala is coordinated by two oxygen atoms in the group C=O and C-O as shown in figure 2. The complex of $[CuL^1_2]$, L^1=2-Hydroxy salicyl hydrazide isatin hydrazine has octahedral geometry [8]. L^1 coordinated to Cu (II) is trident through the group C=O, C=N and C-O as shown in figure 3. The complex of $[CuCl_2(PABA)_2]$ (PABA = p-amino benzoic acid) is a square planar geometry [9] with two $-NH_2$ and two Cl^- ions are coordinated to Cu(II) ion as shown in figure 4.

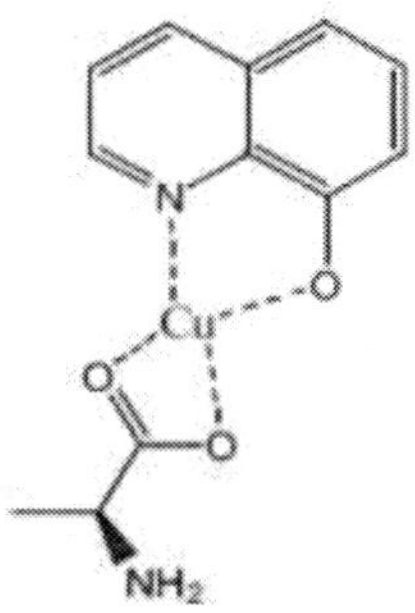

Figure 2. Complex of [Cu(HQ)(ala)] which has tetrahedral geometry [7].

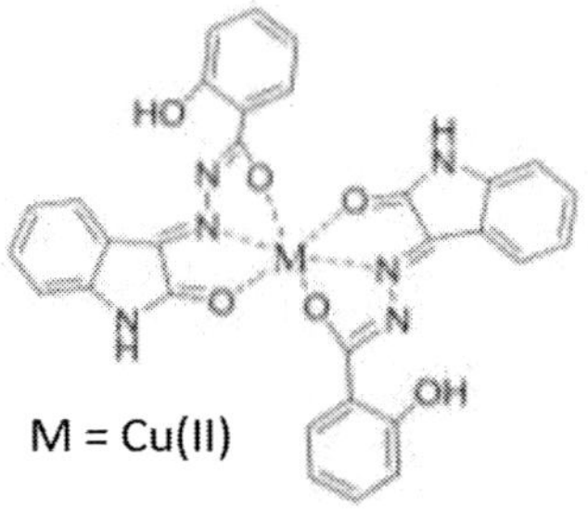

Figure 3. Complex of $[CuL^1_2]$ which has octahedral geometry [8].

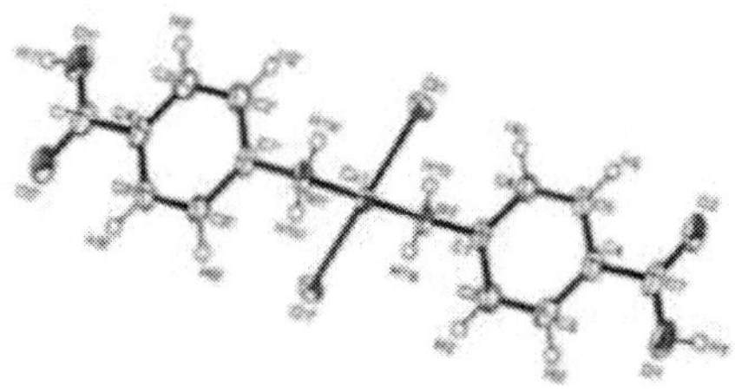

Figure 4. Complex of $[CuCl_2(PABA)_2]$ which has square planar.

Some examples above show that the Cu(II) can be coordinated with various functional groups such as $-NH_2$, C=O, C-O, Cl and H_2O as well as producing a different geometry. Benz has a functional group that can be coordinated to Cu(II) to form complex compounds with a specific geometry. The complex of Cu(II) and benz are expected to enhance the antibacterial activity.

2. Experimental

2.1 Materials

All chemicals, except benz, were purchased from E.Merck and used without further purification, benz were purchased from Kimia Farma.

2.2 Physical measurements

The copper content was determined by Atomic Absorption Spectrometer (AAS) Shimadzu AA-6650. IR spectra were recorded on a Spectrophotometers Prestige-21 Shimadzu. The thermogravimetric analysis of the metal complex was performed by a TG/DTA Diamond Perkin Elmer analyzer. Spectra UV-VIS was performed with Shimadzu UV-3601 spectrophotometer. Molar conductivity (Λ*m) of 1 mM solution in methanol was measured at 25 °C with a Jenway CE 4071 Conductivity meter. The magnetic moment was measured with Auto Sherwood Scientific 10169 Magnetic Susceptibility Balance.

2.3 Determination of coordination numbers of Cu(II)

$CuCl_2 \cdot 2H_2O$ 0,02 M and benz 0,1 M were dissolved in ethanol then mixed with a mole ratio of Cu(II):benz =1:0 to 1:7. The absorbance of all mixed solution was measured using UV-Vis spectrophotometer and graphed maximum wavelength versus mole ratio to determine optimum mole ratio of Cu(II) and benz.

2.4 Synthesis of Cu(II) complex

$CuCl_2 \cdot 2H_2O$ (0.51 g; 3 mmol) was dissolved in ethanol (15 mL) then dropwise into a solution of benz (1.98 g; 12 mmol) in ethanol (35 mL) with stirring, the stirring was continued for 30 minutes then allowed to stand for 30 minutes. The precipitate was filtered off and then recrystallized with methanol

(50 mL). The solution was concentrated to 25 mL then allowed to stand overnight. The precipitate was filtered and dried under vacuum (1.84 g; 91.96 %).

2.5 Antibacterial screening
Antibacterial activities of Cu(II), benz, and the synthesized complex were determined against *Staphylococcus aureus* and *Escherichia coli* by the modified Kirby-Bauer method. The test solution benz, $CuCl_4.2H_2O$ and Cu(II)-benz were prepared in the concentration (% m/v): 10, 25, 50, 75 and 100 in DMSO. Negative controls carried out on the solvent DMSO and the positive were tested against mentioned organisms. Test bacteria cultures were performed in a nutrient agar media on Petri dishes. The nutrient media was prepared with usual components and properly sterilized in an autoclave. Activity was determined by measuring the diameter of zone showing complete inhibition and has been expressed in mm.

3. Result and Discussion

3.1 Determination of coordination numbers of Cu(II)
Cu(II) mixed with benz shifted maximum wavelength into smaller wavelength to achieve the stability, that was when it reached the maximum number of moles benz able to be coordinated with one mole of Cu(II). This was obtained from the intersection graph of λ_{max} (nm) and the mole ratio of Cu (II): benz. Figure 5 shows the intersection of the lines in a mole ratio of metal to ligand 1: 3, this implies that the most optimum mole ratio of Cu(II) and benz is 1: 3.

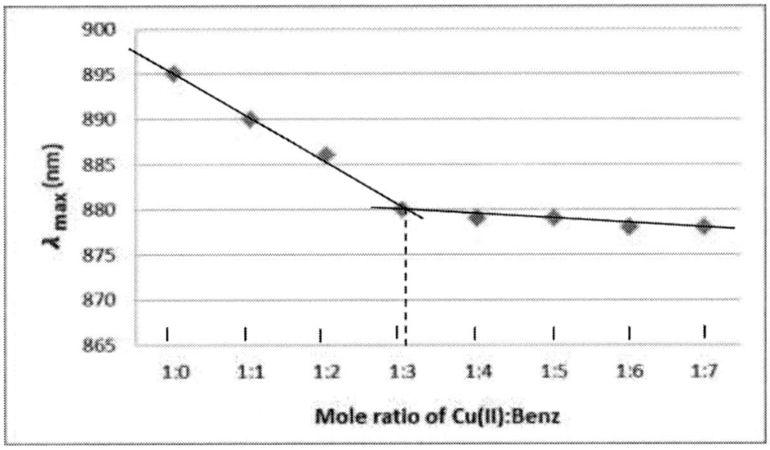

Figure 5. Graphic λ_{max} versus the mole ratio of Cu(II) and benz.

3.2 Synthesis of Cu(II) complex
A mixture solution of $CuCl_2{\cdot}2H_2O$ in ethanol to a solution benz in ethanol produced brown precipitates. The formation of complex compounds was indicated by a color change of the solution from colorless to brown. The shift of λ_{max} was 25 nm, shown in figure 6, indicating the formation of complex Cu(II) Benz. It occurred also in the synthesis of complex Cu(II)-morpholine as well, the complex was indicated due to 20.3 nm shift of λ_{max} towards smaller [10].

3.3 Analysis of the Amount of Copper with Atomic Absorption Spectroscopy
Measurements of copper in the complex was expected to determine the complex formula by comparing the measurement experimental results and theory. The measurement of copper content in the complex based on the experiment was 9.37 ± 0.19%, therefore the estimated complex formula was $Cu(benz)_3(H_2O)_2Cl_2$ or $Cu(benz)_3(H_2O)_3Cl_2$.

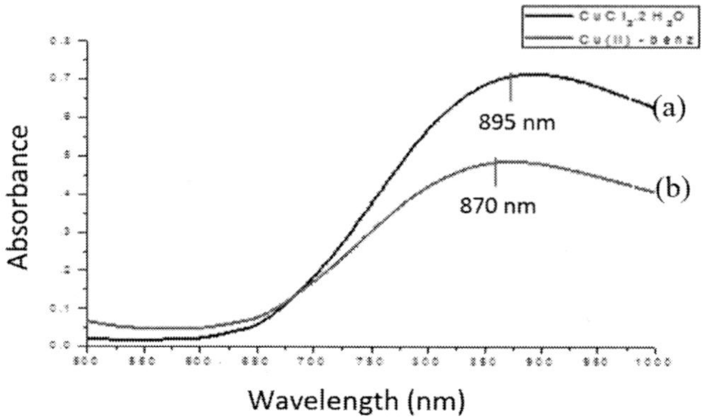

Figure 6. Electronic spectra of $CuCl_2{\cdot}2H_2O$ solution (a) and Cu(II)-benz solution (b).

3.4 Thermal analysis by TG/DTA

Thermal analysis in the complex Cu(II)-benz was used for estimating the presence of water molecules. It reveals that the stability of the complex reaches the temperature of 150°C. After that, the endothermic reaction occurs in the area of 177°C followed by the reducing mass of the complex in the area of 151-188°C. This shows that at these temperatures, as much as 5.54% mass reduction occurred equivalent to the loss of two molecules of H_2O. The temperature of 150-200 °C is a range of coordinated water molecules loss [11]. The complex Cu(II) was estimated to contain three molecules of benz and two molecules of coordinated water. Thus, estimated formula was $Cu(benz)_3(H_2O)_2Cl_2$.

3.5 Analysis of chloride ion

Qualitative ion analysis was used for determining the position of the chloride ion in the complex as a ligand or a counter ion. The analysis was performed by adding a solution of $AgNO_3$ to a solution of each complex in distilled water. The results showed white precipitate of AgCl indicating that chloride acted as a counter ion according to the reaction:

$$Cl^-_{(aq)} + AgNO_{3\,(aq)} \rightarrow AgCl_{(s)} + NO_3^-{}_{(aq)}$$

Ion analysis results showed a white precipitate when the $AgNO_3$ solution was added to a solution of Cu(II)-benz. This shows that there were chloride ions that acted as a counter ion. However, according to the results of AAS in the complex Cu(II)-benz there were two chloride ions, therefore the number of chloride ion as a counter ion was determined by analysis of electrical conductivity.

3.6 Analysis of electrical conductivity

Analysis of electrical conductivity was used to determine the ratio of cations and anions in the Cu(II)-benz by comparison with standard solutions. The molar conductance values of the complex lie in 5,57 ± 0,06 $S.cm^2.mol^{-1}$, thus indicating that is an electrolyte. By comparing the conductivity values of the complex sample with a molar conductivity of the standard solution, it can be seen that the charge ratio of cationic: anions= 1:1. This suggests that one of the Cl^- ion was coordinated to the central metal ion Cu(II) and another acted as a counter ion. Thus, the most probable formula of the complex is $[Cu(benz)_3(H_2O)_2Cl]Cl$.

3.7 Infrared analysis

Infrared (IR) analysis was used for estimating the group coordinated to Cu(II). Absorption band of -NH [12] on the complex (3303 and 3158 cm^{-1}) shifts towards smaller values compared to that of benz (3422 and 3342 cm^{-1}), indicating primary amine coordinated to central metal ion. N-H in Cu(II)-benz has smaller wavenumber than the free ligand due to limited N-H vibration. Thus, it is estimated that benz was coordinated to Cu(II) through primary amine group. It strengthens with the absorption band of Cu-N in 543 cm^{-1} [5]. In $CuCl_2{\cdot}2H_2O$, IR spectra exhibited strong absorption band of O-H in region 3482 cm-[1] [13], while the IR spectra of Cu(II)-benz reveals weak absorption of O-H group appearing in region 3398 cm^{-1}. The overlap between the absorption of O-H and N-H might happen due to an adjacent area, thus O-H absorption hardly appeared in the spectra of Cu(II)-benz. However, the absorption band of Cu-O in region 687 cm^{-1} [5] strengthen the possibility of water molecules coordination in the complex compound.

3.8 Magnetic properties
The magnetic moment of $[Cu(benz)_3(H_2O)_2Cl]Cl$ (μ_{eff}) was 1.98 ± 0.03 BM. It indicated that the compound was paramagnetic with one unpaired electron [14], [15], [16]. It also showed that the Cu(II) configuration was d^9 [17]. The effective magnetic moment was in range 1.82 to 2.32 BM owned by Cu(II) complex with six coordination number and octahedral [18], [19], [20]. In general, Jahn-Teller distortion frequently occurred in the complex of Cu(II) with configuration d^9 [11], [21], [22]. Cu(II)-benz. The complex was formed by coordination of a primary amine group, Cl^- ions, and water molecules to Cu(II). Each adjacent atom will produce a repulsion, thus each different atom will produce different strength of repulsion. The distance and position of each atom will be changed to achieve the least possible repercussion. These circumstances affected the bond lengths between atoms. In the complex of Cu(II)-benz, Cu(II) was bonded to six different groups and there was a possibility of distortion in the octahedral geometry.

3.9 Electronic spectra
Electronic spectra analysis was conducted to determine maximum wavelength, molar absorptivity, and estimated electron transition. It appears that $[Cu(benz)_3(H_2O)_2Cl]Cl$ has a maximum wavelength in region 870 nm (11494.25 cm^{-1}) and a molar absorptivity of 241.96 L/mol.cm. Figure 6 shows a wide absorption in the visible region (870 nm) on $[Cu(benz)_3(H_2O)_2Cl]Cl$ and the possibility of the complex octahedral geometries. Jahn-Teller Distortion caused splitting in the orbit and the result were three transitions B1G→A1g (ν1), B1G→B2G (ν2), B1G→ Eg (ν3) [23],[24] which would appear as three peaks in the electronic spectra. However, the three transitions in the energy level that was close together so often appeared as a wide absorption band [25], [26]. Thus, the suggested geometry of $[Cu(benz)_3(H_2O)_2Cl]Cl$ was distorted octahedral. The suggested structure of the complex is distorted octahedral geometry as shown in figure 6. According to IUPAC nomenclature, the name of the complex is diaquachloro-tris(ethyl-p-aminobenzoate)copper(II) Chloride.

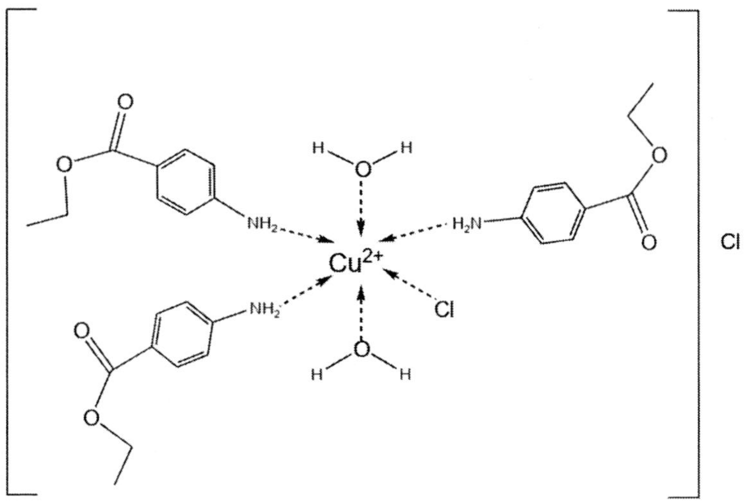

Figure 7. Suggested structure of $[Cu(benz)_3(H_2O)_2Cl]Cl$.

3.9 Antibacterial screening

The test results in the form of antibacterial activity inhibition zone diameter (mm) against Staphylococcus aureus and Escherichia coli are shown in table 1. Table 1 shows that $[Cu(benz)_3(H_2O)_2Cl]Cl$ has better antibacterial activity than Cu(II) and benz. This can be explained by the overtone concept and Tweedy's chelation theory. According to the overtone concept, the entrance to the cell is surrounded by lipid membranes that can only be passed by lipid-soluble materials, therefore liposolubility is an important factor in controlling antimicrobial activity [16]. Meanwhile, according to Tweedy's chelation theory, the polarity of Cu(II) is reduced due to their orbital overlap thus sharing positive charge with the atom donor [8], [27], [28]. This can cause the hydrophobic and lipophilic character and help increase permeability to the lipid membrane. Increased penetration can interfere cellular respiration process and facilitate entry into the cell. The complex can inhibit protein synthesis in normal cells and can ultimately inhibit bacterial growth.

Table 1. Average Diameter of Inhibition Zone of $CuCl_2{\cdot}2H_2O$, $[Cu(benz)_3(H_2O)_2Cl]Cl$ and Benz against *Staphylococcus aureus* and *Escherichia coli.*

Sample	Concentration (m/v)	Average Diameter of Inhibition Zone (mm)					
		Staphylococcus aureus			*Echerichia coli*		
		24 h	48 h	72 h	24 h	48 h	72 h
$CuCl_2{\cdot}2H_2O$	5%	0,00	0,00	0,00	0,00	0,00	0,00
	10%	2,23	4,23	4,70	0,00	0,00	0,00
	25%	5,27	6,87	7,10	4,57	6,70	7,10
	50%	8,53	9,83	10,13	7,23	9,77	10,17
	75%	12,63	15,40	15,53	12,13	14,20	14,60
	100%	14,77	17,93	18,67	19,10	21,10	22,30
Benz	5%	0,00	0,00	0,00	0,00	0,00	0,00
	10%	1,43	2,90	3,43	1,43	2,67	3,37
	25%	3,20	3,93	4,10	3,27	3,80	4,07
	50%	4,73	6,03	6,37	6,23	8,00	8,27
	75%	8,20	9,90	10,47	8,03	11,30	11,57
	100%	10,77	14,13	14,73	12,00	15,50	16,03
Cu(II)-benz	5%	1,23	1,83	2,20	0,83	1,50	1,77
	10%	4,03	4,77	5,03	2,50	4,47	4,97
	25%	6,80	8,43	8,33	6,83	9,20	9,47
	50%	11,50	13,73	14,57	9,83	13,07	13,30
	75%	15,67	18,27	19,03	15,83	17,87	18,93
	100%	20,13	24,27	24,87	23,10	26,20	26,40

4. Conclusion

The complex $[Cu(benz)_3(H_2O)_2Cl]Cl$ was able to be synthesized by mixing solution of Cu(II) in ethanol and benz solution in ethanol with a mole ratio of 1: 4 and stirred for 30 minutes. Functional groups coordinated to the Cu(II) were possibly NH_2, H_2O, and Cl^- ions forming a distorted octahedral geometry. The complex $[Cu(benz)_3(H_2O)_2Cl]Cl$ increased antibacterial activity significantly compared $CuCl_2{\cdot}2H_2O$ and benz.

Acknowledgement
The authors gratefully express acknowledgments to Indonesian Directorate General of Higher Education (DIKTI) for financial support.

References

[1] Fulias A, Vlase G, Grigorie C, Ledeti, I, Albu P, Bilanin M, and Vlase T 2013 *J Therm Anal Calorim* **113** 265–271
[2] Paun A, Zarafu I, Caproiu M T, Draghici C, Maganu M, Cotar, A I, Chifiriuc, M C and Ionita P 2013 *Comptes Rendus Chimie* **16** 665–671.
[3] Devi J and Batra N 2015 *Spectrochim Acta A* **135** 710–719
[4] Yimer A M., 2015 *Review of Catalysts* **2(1)** 14-25
[5] Liu H, Yang W, Zhou W, Xhu Y, Xie J and Li M 2013 *Inorg. Chim. Acta 405* 387–39
[6] Katwal R, Kaur H and Kapur B K 2013 *Sci. Rev. Chem. Commun.* **3(1)** 1-15
[7] Amolegbe S A, Adewuyi S, Akinremi C A, Adediji J F, Lawal A, Atayese A O, Obaleye A J 2014 *Arabian J. Chem. Article in press* DOI: 10.1016/j.arabjc.2014.11.040
[8] Jamil W, Solangi S, Ali M, Khan K M, Taha M and Khuhawar M Y 2015 *Arabian J. Chem. Article in Press,* DOI : 10.1016/j.arabjc.2015.02.015
[9] Guedes G P, Farias F F, Novak M A, Machado F L A and Vaz, MGF 2011 *Inorg. Chim. Acta* **378** 134–139
[10] Rahadjo S B, Wahyuningsih S and Atmojo S 2005 *Alchemy* **4(2)** 67-74
[11] Abdelkarim, A T 2015 *I. J. Pharm. Sci.* **5(1)** 839-851
[12] Inci D, Aydın R, Yılmaz, Dilek, Genckal H M, Vatan O, Cinkılıc N and Zorlu Y 2015 *Spectrochim. Acta A* **136** 761–770
[13] Mistri S, Bertolasi V and Manna S C 2015 *Polyhedron* **88** 101–109
[14] Chandraleka S and Chandramohan G 2014 *African Journal of Pure and Applied Chemistry* **8(10)** 162-175
[15] Mahal A, Abu-El-Halawa R, Zabin S A, Ibrahim M, Al-Refai M, Kaimari T 2015 *World Journal of Organic Chemistry* **3(1)** 1-8
[16] Gupta O P and Kumar A 2015 *Journal of Chemistry and Chemical Sciences* **5(2)** 69-78
[17] Pahonțu E, Ilieş D, Shova S, Paraschivescu C, Badea M, Gulea A and Roşu T 2015 *Molecules* **20** 5771-5792.
[18] Kumar S M, Dhahagani K, Rajesh J, Anitha K, Chakkaravarthi G, Kanakachalam K, Marappan M and Rajagopal G 2015 *Polyhedron* **85** 830–840
[19] Pechiammal M, Singh L, Ravichandran S 2015 *International Journal of ChemTech Research* **7(01)** 287-292.
[20] Venugopal K, Rameshbabu K and Sreeramulu J 2015 *Der Pharma Chemica* **7(1)** 252-260.
[21] Batra N and Devi J 2015 *Journal of Chemical and Pharmaceutical Research* **7(2)** 183-189
[22] Devi J and Batra N 2015 *Spectrochim Acta A* **135** 710–719
[23] Dessie B K, Zenebe M M and Assefa A G 2015 *American Journal of Applied Chemistry* **3(2)** 40-47
[24] El-Gamel NEA and Zayed M A 2015 *World Journal Of Pharmaceutical Research* **4(01)** 142-162
[25] Ciolan F, Patron L, Marutescu L, Carmen M and Chifiriuc 2015 *Farmacia* **63** 1
[26] Nagaraj K, Sakthinathan S and Arunachalam S, 2015 *Journal of Iran Chemistry Society* **12** 267–275
[27] Moosun, S B, Jhaumeer-Laulloo, Hosten E C, Betz and Bhowon M G 2015 *Inorg Chim Acta* **430** 8–16
[28] Patel M N, Joshi H N and Patel C R 2013 *Spectrochim Acta A* **104** 48–55.

Smart Natural Fiber Reinforced Plastic (NFRP) Composites Based On Recycled Polypropylene in The Presence Kaolin

N S Suharty[1], H Ismail[2], K Diharjo[3], D S Handayani[1] and W A Lestari[1]
[1] Chemistry Department, Faculty of Mathematics and Natural Sciences, Sebelas Maret University, Jl. Ir Sutami 36A Surakarta, Indonesia
[2] School of Materials and Mineral Resources Engineering, Engineering Campus, University Sains Malaysia, 14300 Nibong Tebal, Pulau Pinang, Malaysia
[3] Faculty of Engineering, Sebelas Maret University, Surakarta, Indonesia
E-mail: nsuharty@yahoo.com

Abstract. Composites contain double filler material which act as reinforcement and flame retardants of recycled polypropylene (rPP)/kaolin(Kao)/palm oil empty bunch fiber (PEBF) have been succesfully prepared. The composites were synthesized through reactively solution method, using coupling agent PP-g-AA and compatibilizer DVB. The effect of double filler [Kao/PEBF] were investigated flexural strength (FS), inflammability, and morphology. Mechanical testing result in accordance to ASTM D790, the FS of rPP/DVB/PP-g-AA/Kao+ZB/PEBF composite was 48% higher than that of rPP matrix. Moreover, flexural modulus (FM) was significantly improved by 56% as compared to that of rPP matrix. The scanning electron images (SEM) shown good dispersion of [Ka/PEBF] and good filler-matrix interaction. The inflammability testing result which is tested using ASTM D635, showed that the flame resistance of rPP/DVB/PP-g-AA/Kao+ZB/PEBF composite was improve by increasing of time to ignition (TTI) about 857% and burning rate (BR) decreasing to 66% compared to the raw material rPP matrix. In the same time, the addition of 20% (w/w) PEBF as a second filler to form rPP/DVB/PP-g-AA/Kao+ZB/PEBF composites (F5) is able to increase: the FS by 17.5%, the FM by 19%, the TTI by 7.6% and the BR by 3.7% compared to the composite without PEBF (F2).

1. Introduction

Today, the automotive production is increasing, however, availability of metal as raw material for the automotive is also depleting [1]. In addition, the fuel oil is a material supporting automotive operations, its availability is also going decrease significantly [2]. This problems can be overcome by the use of light weight automotive. The light weight automotive can be created by partially replacing metal in automotive by composite materials [3]. Composite is a very interesting material by nature to produce new materials with new properties as required. Manufacture of high mechanical properties composite can be done by adding a clay or fiber as reinforcement in polymer matrix. One alternative development composites is by creating smart composite Natural Fiber Reinforced Plastic (NFRP) which has advantages such as light in weight, strong and biodegradable [4,5]. Polypropylene is a plastic used widely in various applications such as electronic devices, transportation and packaging [6,7]. Polypropylene in the packaging of mineral water waste overflow and pollute the environment, it can be used as a polymer matrix [8,9]. The use of fiber as reinforcing filler for plastic composites have been used extensively. Maulida [10] synthesized biocomposite of acrylonitrile butadiene styrene (ABS) with PEBF (palm oil empty fruit bunch fiber) was obtained composites have high mechanical properties. Wirjosentono [11] synthesized bio-composites of polypropylene with PEBF was obtained composites have high mechanical properties. Other reinforcing material that has been used by previous researchers is clays [12-15]. Any clay other than as a reinforcing material can also act as flame retardants. Shehata [16] synthesized a composite of polypropylene and kaolin clay to produce a composite that has high flame resistance. Du [17] make a composite of polypropylene and halloysite clays produce a composite that has high flame resistance.

In this research will synthesized : (1) rPP/DVB/PP-g-AA/Kao+ZB composite by using the rPP as a matrix, divinylbenzene (DVB) as a crosslinker and PP-g-AA as a coupling agent and the kaolin (Kao) as a first reinforcement, (2) rPP/DVB/PP-g-AA/Kao+ZB/PEBF composite containing palm oil empty bunch fiber (PEBF) and Kaolin has used as a double reinforcement filler. The composites in the form double filler PEBF and Kaolin, which produces NFRP composites with mechanical properties and durability of high flame and biodegradable. GeoBioComposites are the development of green composites that fulfill the criteria to be applied in automotive components.

2. Methods

2.1Materials

Recycled polypropylene (rPP) was collected from plastic recycling centre in Surakarta Indonesia. The rPP prepared by cleaning up the waste cup and then chopped approximately in 2 mm x 2 mm. PEBF obtained from oil palm plantations in Lampung Indonesia. The PEBF was then alkalized with 4% NaOH for one night, and washed with distilled water and then dried in an oven at 60 °C before grounded into 200 mesh sized particles. The kaolin purchased from PT Brataco in Surakarta Indonesia. It crushed and sieved with a 200 mesh sieve and calcined at temperatures of 800 °C for 1 hour. All chemicals used in this study were pro-analysis without any purification. Other chemical compounds, such as acrylic acid (AA) and divinyl benzene (DVB), xylene derived from Aldrich; while benzoyl peroxide (BPO) and zinc borate (ZB) purchased from E.Merck.

2.2 Composites manufacturing process

Formulations of rPP (F0), rPP/DVB/PP-g-AA/Kao+ZB (F2), rPP/DVB/PP-g-AA/PEBF (F3) and rPP/DVB/PP-g-AA/Kao+ZB/PEBF (F5) composites synthesis are given Table 1. Composite synthesis is reactively done by solution process using xylene, equipped with a condenser, mechanical stirrer and N2 gas. All starting materials: rPP, DVB, PP-g-AA, Kaolin, and PEBF solved in boiled xylene and mixed perfectly for 1 hour. The mixture product was then evaporated to release the xylene to form masterbatches (MB's) composites. The MB's was hot pressed for 20 minutes at 180 °C to produce the specimen for mechanical test [8].

Table 1. Formulation of rPP/DVB/PP-g-AA/ Kaolin/PEBF composites with and without ZB

Ingredients	Composites					
(phr)	F0	F1	F2	F3	F4	F5
rPP	100	80	80	80	80	80
PEBF	-	-	-	20	20	20
Kaolin	-	20	20	-	20	20
ZB	-	-	5	-	-	5

2.3 Testing of materials

The testing of Flexural Strength (FS) and Flexural Modulus (FM) was carried out using universal testing machine (UTM) by following ASTM D790. The testing of flame resilience of composites in accordance with ASTM D635 to determine the value of time to ignition (TTI) and burning rate (BR).

2.4 Morphology analysing

The fracture surface of specimens after FS testing were characterized using SEM-EDX JEOL JSM-6360 LA to analyse their morphology, especially on the cross section area the fracture surface.

3. Results and Discussion

3.1 Mechanical properties of composites

The mechanical properties of composites are presented in Figure 1-4. The FS and FM of rPP (F0) as raw materials was 23.61 MPa and 790 MPa. The FS of rPP/Kao+ZB (F2) enhanced 16.5% (27.5 MPa) and the FM also increased 22% (960 MPa) due to the effect of the adding 20% (w/w) kaolin as a first reinforcement. Meanwhile, the presence 20% (w/w) PEBF on rPP (F3) produce rPP/PEBF composite with 30.11 MPa of FS increased 23% and 1060 MPa of FM also increased 29% compared to rPP. In the addition of 20% (w/w) kaolin as a first reinforcement and 20% (w/w) PEBF as a second reinforcement (F5), the FS of rPP/Kao+ZB/PEBF composites become 32.3 MPA in FS and 1140 MPa in FM, or the FS increased 48% and the FM increased 56% compared to that of rPP (Figure 1, 2, 3 and 4). The increases was caused by both PEBF and kaolin act as a reinforcing material [14,18,19].

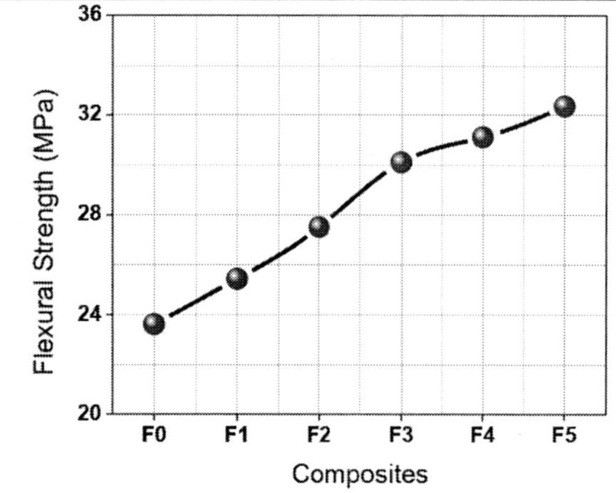

Figure 1. Flexural Strength of the composites (MPa)

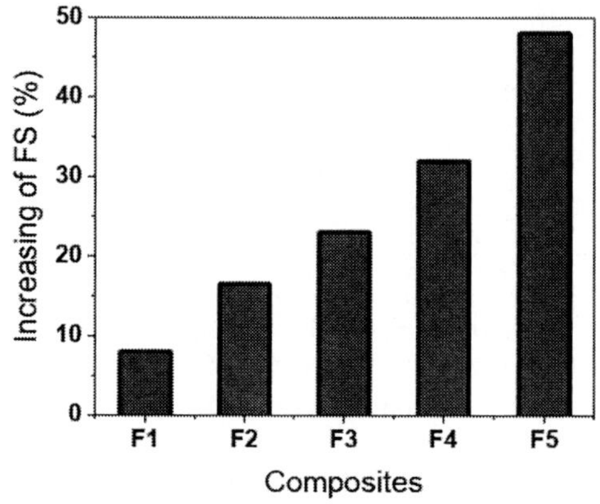

Figure 2. Comparison of Increasing FS to that of rPP (%)

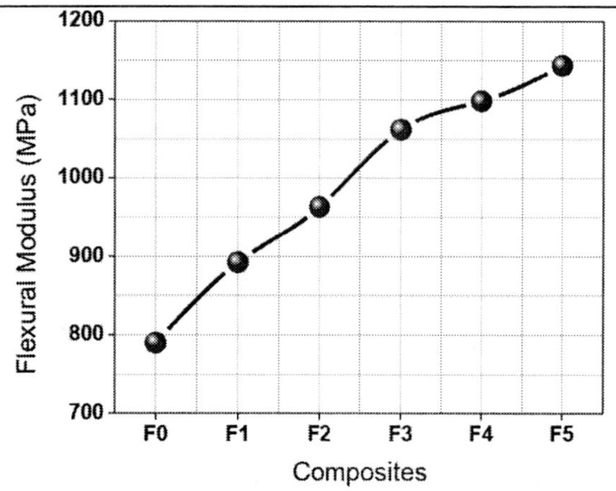

Figure 3. Flexural Modulus (FM) of the composites

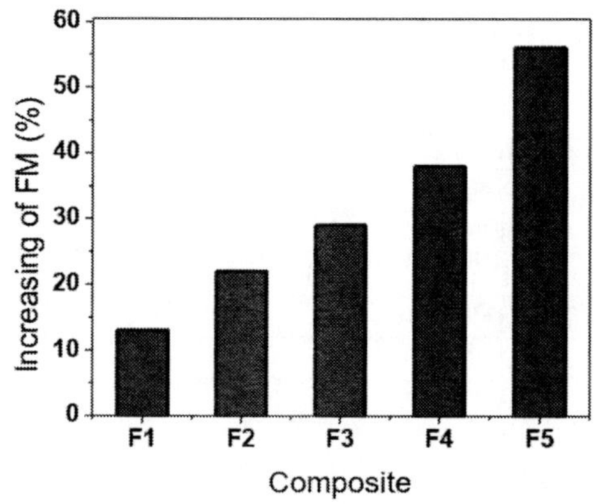

Figure 4. Comparison of Increasing FM to that of rPP (%)

The existence of a reinforcing material in the form of fibers within the polymer matrix serves to hold the burden that has been transferred from the matrix to the fiber. Chemically, PEBF have polar groups such as -OH are able to interact chemically with metin group of RPP through a coupling compound PP-g-AA [7]. Furthermore, the presence of kaolin clay reinforcing material that has a layer of type 1: 1 can improve the mechanical properties. It is caused by the geometry of the clay in the polymer matrix can be distributed both intercalation and exfoliation [20,21]. Judging from the chemical side, kaolin clay has a layer in the form of Si-O tetrahedral and octahedral layers of Al-OH which are capable of interacting with polar groups on a polypropylene metin through the coupling compound PP-g-AA. Maximum interaction between filler and kaolin PEBF with PP matrix will lead to increased mechanical properties [7].

3.2 Inflammability properties

To prove the inflammability of various (F0-F5) composites its performed resistance combustion testing. The method horizontal burning of ASTM D635 is used for testing composites. The result is presented by the time to ignition (TTI) and the burning rate (BR), which can be seen in Figure 5-8. The TTI and BR of rPP (F0) was 1.02 second and 19 mm/mins, and in the presences of 20% (w/w) kaolin (F2), the TTI of rPP/kaolin+ZB increased 806% (9.2 second), however, the BR decreased 48% (9.8 second) compared to rPP. The increased of TTI and the decreased of BR value its due to the effect of 20% kaolin + 5% ZB. When kaolin + ZB burned, its will form the charcoal and overlying the surface of the specimen, thus limiting the supply of oxygen. This condition can inhibit the TTI and slow down its BR. The presence of 20% (w/w) PEBF (F5) in rPP/kao+ZB/PEBF, the TTI increased to 857% (9.9 seconds) and the BR decreased to 66% (7.8 mm/mins) compared to that of rPP. Cellulose fibers of PEBF during the combustion process formed charcoal as a barrier that prevents the entry of oxygen at the surface. Kaolin undergo endothermic reaction which can absorb heat energy during the combustion process takes place, while ZB can release water vapor to cool the system [19,22,23]. The synergy of these three components causing the flame resistance of the composite rPP/DVB/PP-g-AA/Kao+ZB/ PEBF (F5) increased significantly. The presence of 5% ZB as fire retardant additive can inhibited TTI and also slow down the BR. The ZB compound binds some water when its burnt will release the water vapour. ZB if burned is also formed B2O3 which is moist and form a stable charcoal. Water that is released by ZB during combustion can cause charcoal to form a foam that inhibits combustion [18,22,23].

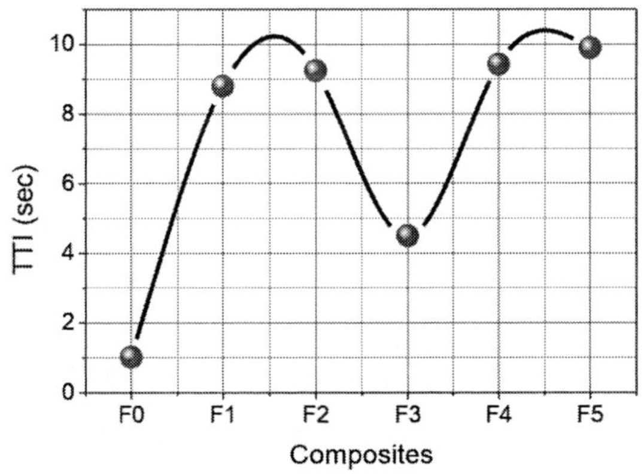

Figure 5. TTI of the various composites

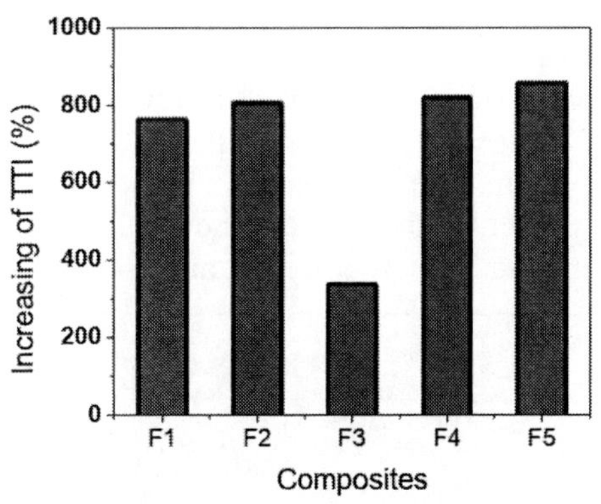

Figure 6. Comparison of Increasing TTI to that of rPP (%)

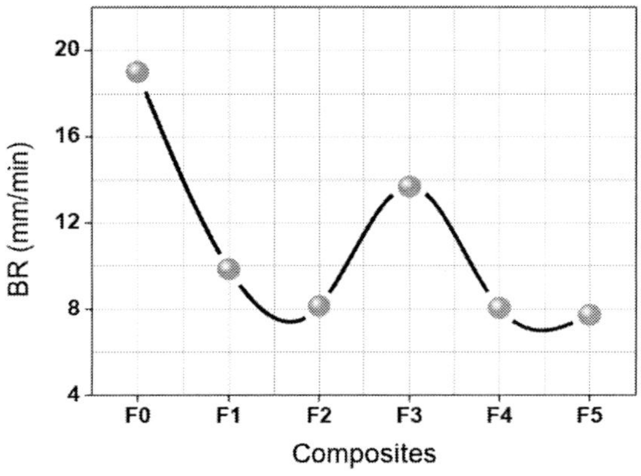

Figure 7. BR of various composites

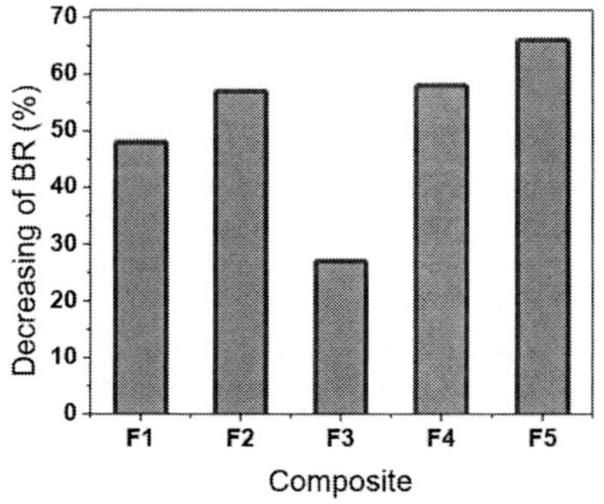

Figure 8. Comparison of Increasing BR to that of rPP (%)

The mechanical and inflammability properties values of rPP/DVB/PP-g-AA/Kao+ZB: the FS is 27.5 MPa, FM is 960 MPa, TTI is 9.2 seconds and BR is 8.1 mm/mins. The effect of mechanical and inflammability properties values by adding 20% (w/w) PEBF is as follow: the FS values increased 17.5%, the FM values increased 19%, the TTI values increased 7.6% and the BR values decreased 3.7% compared to rPP/DVB/PP-g-AA/Kao+ZB composites without PEBF.

3.3 Morphologies analysis

The fracture surface of specimens after FS testing were characterized using SEM image especially on the cross section area of the fracture surface. The morphological study carried out by compare the composite cross-section SEM image of the fracture of the specimen testing FS rPP/Kao without and with PEBF, can be seen at Figure 9 (F2) and Figure 10 (F5). The observation of SEM image F2 shown that the kaolin particles as seen at points A very well dispersed in the PP matrix. However, at point B, the result of fracturing after FS testing showed the brittle nature of the material without PEBF (F2), its mean the interaction between filler and matrix is good. Morphology of composite rPP/Kao/PEBF (F5) can be seen in the SEM image on points D and E, both filler kaolin and PEBF dispersed very well. And also at points C after FS test shown the ductile composite nature. This means there has been a very strong interaction between filler and matrix, resulting a strong mechanical properties of the composites.

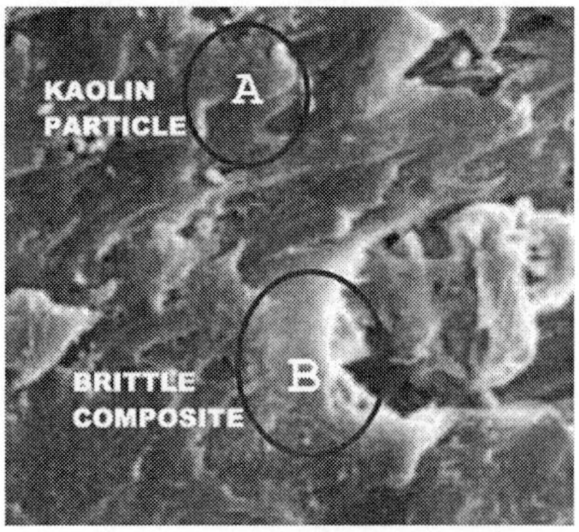

Figure 9. SEM of PP/DVB/PP-g-AA/Kao (F2) after Fleural Strength (FS) Testing

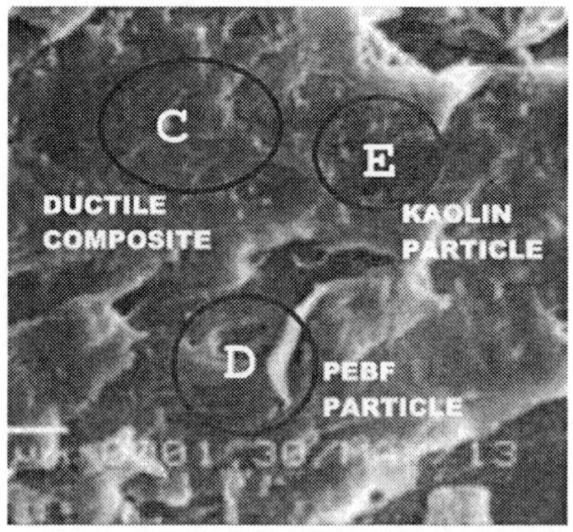

Figure 10. SEM of PP/DVB/PP-g-AA/PEBF /Kao/ZB (F5) after FS Testing

4. Conclusion

The mechanical and inflammability properties values of rPP/DVB/PP-g-AA/Kao+ZB is as follow: FS is 27.5 MPa, FM is 960 MPa, TTI is 9.2 seconds and BR is 8.1 mm/mins. The effect of mechanical and inflammability properties values by adding 20% PEBF is as follow: the FS values increased 48%, the FM values increased 56%, the TTI values increased 857% and the BR values decreased 66% compared to rPP as raw material. In the same time, the effect of mechanical and inflammability properties values by adding 20% PEBF is as follow: the FS values increased 17.5%, the FM values increased 19%, the TTI values increased 7.6% and the BR values decreased 3.7% compared to rPP/DVB/PP-g-AA/Kao+ZB composites without PEBF.

Acknowledgment

The author would like to thank for financial support in this project through International Research Collaboration and Scientific Publication Grants to Directorate General of Higher Education Indonesia.

References

[1] Gordon R B, Bertram M, and Graedel T E 2006 *Proceeding of the National Acedemy of Sciences,* **103** 1209-1214
[2] Sorrell S, Speirs J, Bentley R, Brandt A, and Miller R 2010 *Energy Policy* **38** 5290-5295
[3] Rezaei F, Yunus R, Ibrahim N A, and Mahdi E S 2008 *Polymer-Plastic Technology and Engineering* **47** 351-357
[4] Arbelaiz A, Cantero G, Fernande B, Mondragon I, Ganan P, Kenny J M 2005 *Polym. Compos.* **26** 324–332
[5] Faruk O, Bledzki A K, Fink H P, and Sain M 2012 *Prog. Polym. Sci* **37** 1552-1596
[6] Li M, Wen X, Liu J, Tang T 2014 *Composites:Part A* **67** 212-220
[7] Suharty N S, Ismail H, Diharjo K, Handayani D S, and Firdaus M 2016 *Procedia Chem.* **19** 253-258
[8] Suharty N S, Wirjosentono B, Maulidan F, Handayani, D S, Sholikhah J, and Maharani Y A 2008 *J. Phys. Sci.* **19** 105-115
[9] AlMaadeed M A, Kahraman R, Khanam P N, Madi N 2012 *Mater. Des.* **42** 289-294
[10] Maulida 2008 *J. Eng. Res.* **1** 74-79
[11] Wirjosentono B, Guritno P, and Ismail H 2004 *I. J.Polym. Mater. Polym. Biomater.* **53** 259-306
[12] Suharty N S, Ismail H, Wibowo F R, Handayani D S, Firdaus M, and Lathifah L 2015 *Adv. Mater. Res.* **1105** 56-61
[13] Ishak Z, Kusmono A M, Chow W S, Takaechi T, and Rochmadi 2008 Proceedings of the Polymer Processing Society 24th Annual Meeting PPS 24 (Salerno: Italy)
[14] Pasbakhsh P, Ismail H, Ahmad Fauzi M N, and Abu Bakar A 2009 *Polym. Test.* **28** 548-559
[15] Prashantha K, Lacrampe M F, and Krawczak P 2011 *Express Polym. Lett.* **5** 295-307
[16] Shehata A B, Hassan MA, and Darwish N A 2004 *J. Appl. Polym. Sci.* **92** 3119-25
[17] Du M, Guo B, and Jia D 2006 *Eur. Polym. J.* **42** 1362-1369
[18] Sain M, Balatinecz J, and Law S 2000 *J. Appl. Polym. Sci.* **77** 260
[19] Al-Hassany Z, Genovese A, and Shanks R A 2010 *J. Polym. Lett.* **4** 70-93
[20] Handge U A, Hedicke-Hochstotter K, and Altstadt V 2010 *Polymer* **51** 2690-2699
[21] Gorrasi G, Senatore V, Vigliotta G, Belviso S, and Pucciariello R 2014 *Eur. Polym. J.* **61** 145-156
[22] Suharty N S, Ismail H, Wibowo F R, Dihardjo K, Handayani D S, Firdaus M, and Lathifah L 2015 *App. Mech. Mater* **736** 13-18
[23] Delhom C D, Ghoorahoo W, dan Pang S S 2010 *Composites: Part B* **41** 475-481

Effects of PVDF concentration on the properties of PVDF membranes

E Pramono[1], A L Simamora[1], C L Radiman[1*], D Wahyuningrum[2]

[1]Inorganic and Physical Chemistry Division, Chemistry Department- Faculty of Mathematics and Natural Sciences, Institut Teknologi Bandung, Bandung, Indonesia
[2]Organic Chemistry Division, Faculty of Mathematics and Natural Sciences, Institut Teknologi Bandung, Bandung, Indonesia

Email: cynthia@chem.itb.ac.id

Abstract. Polyvinylideneflouride (PVDF) is a good polymeric material for preparing ultrafiltration and microfiltration membranes due to its high mechanical properties and chemical resistance. The objective of this work is to study the effects of PVDF concentration on the membrane properties such as mechanical strength, permeability of water and permselectivity of T-500 and T-2000 dextran solutions. These membranes were also characterized by contact angle determination and its morphology was observed by scanning electron microscopy (SEM). From the experimental data, it can be concluded that PVDF concentration affects the surface properties, permeability and permselectivity of the produced membranes. Higher PVDF concentrations results in higher hydrophobicity, mechanical properties and rejection towards T-500 and T-2000 dextrans, but lower water flux.

1. Introduction

Membrane technology has been applied for ultrafiltration (UF) and microfiltration (MF) since several decades, especially for separation process and water treatment [1-5]. Many researchers are still developing various polymeric membranes in order to improve the filtration performances and efficiency. Polyvinylideneflouride (PVDF) is considered as a promising polymer for UF and MF applications due to its high mechanical properties and chemical resistance [6, 7].

The characteristics of PVDF membranes are affected by various factors such as preparation method, composition of polymer, solvent and additive, porogene and coagulation bath [8-10]. UF and MF membranes are usually prepared by phase inversion method which is quite simple and yields asymmetric membranes with appropriate pore dimension. The addition of additives and porogene influence the diffusion rate of solvent and non-solvent during the phase inversion [11]. Polar additives results in coarser membrane surface, while the additions of porogene in pore dimension [12, 13]. Membrane composition and coagulant temperature affect the resulting membrane structure. The addition of electrolyte or polar organic compound such as glycerol results in membranes having longer microvoid structures, while low temperature coagulant yields faster polymer crystallization [14, 15]. Other researchers found that the use of alcohols in the coagulation bath produced membranes with higher hydrophobicity [16].

In this work, PVDF membranes have been prepared by phase inversion, in which the concentration of polymer was varied. The objective of this work is to study the effects of PVDF concentration on the membrane properties such as mechanical strength, permeability of water and permselectivity of T-500 and T-2000 dextran solutions. These membranes were also characterized by contact angle determination and its morphology was observed by scanning electron microscopy (SEM).

2. Experimental

2.1. Materials

PVDF powder (Solef® 1015, Solvay), polyethylene glycol (PEG)-400 (Merck), dimethylacetamide – DMAc (99%, Merck), H_2SO_4 (96%, Merck), phenol (99%, Merck), dextran T- 500 (M_n 500 kDa) and T-2000 (M_n 2000 kDa) (Pharmacosmos) were used without further purification.

2.2. Membrane preparation

A mixture of PVDF and PEG 400 was put in DMAc and stirred at 60 °C until the polymer was completely dissolved and formed a dope. The concentration of PVDF used was 15, 17, 19 and 21% w/w and the PEG was kept constant at 5% w/w. The dope was cast on a glass plate and directly put in a water coagulant. The resulting membrane was then separated from the plate and washed in running water for several hours in order to remove the residual solvent. The membranes were then kept in glycerin before being characterized.

2.3. Membrane characterization

The contact angle was measured by droplet method using Tantec Contact Angle Meter. The solid membrane was cut according to the dimension of the sample holder and a droplet of water was put on it. The angle was recorded and the data was taken at different points randomly. The morphology of the membranes was observed by using SEM JEOL JSM-6360LA. The PVDF membranes were gold coated and the photos of membrane surfaces and cross-sections were taken using 2000 and 10,000 magnification. The mechanical properties of the membranes, taken as the force at break and elongation at break, were measured by using Tensolab mechanical tester. Membranes of 3 × 50 mm were cut according to ASTM D 3822-07. The data was taken for 5 to 6 samples and the average value was calculated.

2.4. Water flux and rejection towards dextran solutions

Both water flux and rejection were determined using a dead-end cell. The PVDF membranes were cut according to cell dimension and put at the bottom of it which has a diameter of 5 cm. The cell was filled with demineralized water and the applied pressure was 1 bar. Prior to measurements, all membranes were compacted for 30 minutes and the volume of the permeate was measured afterwards. Equation (1) was used for calculating the water flux.

$$J = \frac{V}{A \times t} \tag{1}$$

where J is the water flux ($L.m^{-2}.h^{-1}$), V is the permeate volume (L), A is the membrane area (m^2) and t is the filtration time (h).

For the measurement of rejection of T-500 and T-2000 dextran solutions with a concentration of 1000 ppm, the feed was changed accordingly. The compaction was again done for 30 minutes and at the end of filtration process, the samples of both permeate and retentate were taken for quantitative analysis. Into those samples, a few mL of phenol was added and the concentration of dextran was measured at 490 nm using Spectronic-20. The percentage of rejection toward dextran was calculated by equation (2)

$$R = \left(1 - \frac{C_P}{C_R}\right) \times 100 \tag{2}$$

where R is the rejection percentage, C_p is the concentration of dextran in the permeate (M) and C_R is the concentration of dextran in the retentate (M)

3. Results and discussion

3.1. Properties of PVDF membrane surfaces

One of the important factors determining the membrane performance is its surface properties. The polarity of membrane surface affects the interaction between membrane and components of the feed solution [17]. In this work the properties of membrane surfaces were measured as the contact angles of the membranes with water. Table 1 shows the data of contact angle of the produced PVDF membranes. It shows that higher PVDF concentration resulted in higher contact angle between membrane and water; indicating that the membranes became more hydrophobic. The 15% PVDF membranes has the lowest hydrophobicity indicated by its lowest contact angle (64°), while the 21 % PVDF membrane has the highest one. It can be explained by the fact that PVDF has an intrinsic hydrophobic nature and hence, higher PVDF concentration resulted in lower interaction between membrane and water. This trend is in accordance with previous research [18].

Table 1. Contact angle of PVDF membrane and water

PVDF (%w/w)	Contact angle (°)
15	64
17	70
19	72
21	78

3.2. Mechanical properties of PVDF membranes

Membrane mechanical properties determine the endurance of membranes toward the applied pressure used during its applications. The effects of PVDF membrane composition on its mechanical properties were shown at Table 2. In general, it can be concluded that higher PVDF concentration resulted in higher tensile strength and strain as well. PVDF membranes show an elastic behaviour indicated by the high value of strain which in certain composition reached more than 100%. This was also reported previously by other researchers [10, 19]. Table 2 shows that 15 % PVDF has the lowest stress and strain values compared to other compositions, while the 19 % PVDF has the highest strain (148%). Membrane elasticity is a good indication of long membrane life-time, since it can avoid membrane fracture during filtration processes [11].

Table 2. Effects of PVDF concentration on membrane mechanical properties

PVDF (%)	Stress (MPa)	Strain (%)
15	5.73 ± 2	64.6 ± 21
17	6.83 ± 1	104.6 ± 12
19	8.29 ± 1	148.8 ± 6
21	9.08 ± 1	91.1 ± 41

3.3. Membrane permeability and permselectivity

Membrane permeability and permselectivity were determined by water flux and rejection towards T-500 and T-2000 dextran solutions. Figure 1 shows the effects of PVDF concentration on water flux and dextran rejections. It can be seen that the water flux increased with PVDF concentration. The highest water flux was shown by 15 % PVDF while the lowest was reached by 21 % PVDF membrane. Regarding the contact angle data, the 15 % PVDF membrane shows the highest polarity which explains why it has strong interactions with water and increases the water flux. This fact proves that membrane permeability and permselectivity are influenced by membrane surface and pore properties [20, 21]. Besides that, the permeability is also affected by the density of membrane pores, which the denser membrane, the smaller water fluxes [22, 23]. Figure 1 also shows that dopes with higher PVDF concentration produce denser membranes.

Membrane selectivity was measured by the rejection of dextran molecules which have a definite molecular size. Figure 1 shows that increasing PVDF concentration produced higher rejection towards dextran. The rejection of T-2000 dextran is higher than the T-500 one, because the dimension of T-2000 molecules are higher than the T-500s. As mentioned previously in the discussion of membrane permeability , membrane rejection is also influenced by surface properties as well as pore dimension [24]. The rejection data shows that the rejection percentage of the produced membranes are in the range of 50 and 85 % which means that some solutes still pass across the membranes. Since the polarity of dextran is much higher than the PVDF, one can expect that the interactions among dextran molecules and membrane surface are not optimal. Therefore, the separation of dextran molecules are predominantly determined by the membrane pores. Consequently, membranes with higher PVDF concentration have denser pores which are able to produce a better separation of dextran molecules.

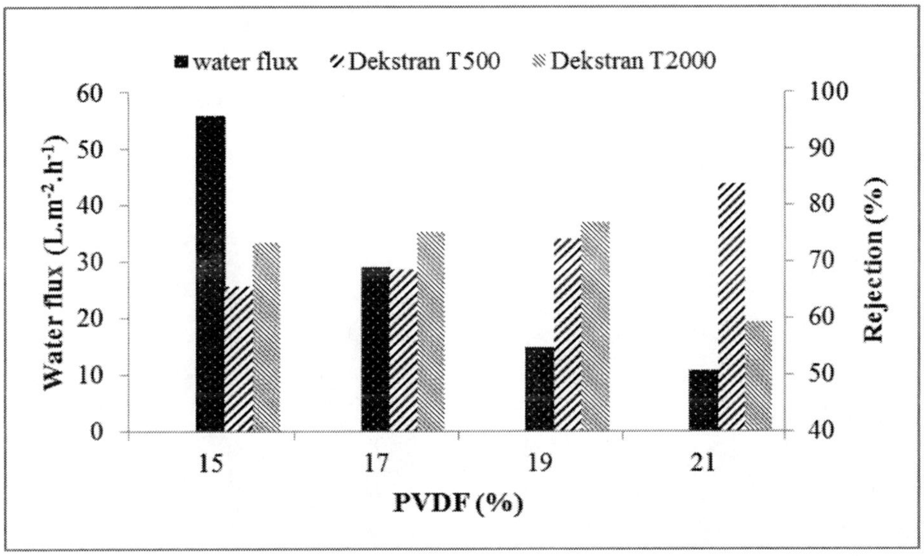

Figure 1. Effects of PVDF concentration on water flux and dextran rejections

3.4. Membrane morphology

Figure 2 and 3 show the morphology of the surface and cross-sections of two types of PVDF membranes, respectively. In general it can be seen that the morphological difference between 15 % and 17 % PVDF membranes are not significant. However, the pore dimensions of the membrane surface are different; 55 to 110 nm on the 15 % PVDF membrane but the pores of the 17 % PVDF membrane could not be measured. The addition of PEG into the dope plays a role of pore former. Dopes with low PVDF have higher ratio of PEG to total polymer concentration and hence, more pores are formed. Figure 2 also shows that membranes with high PVDF concentration result in denser membranes having lower permeability but higher permselectivity.

Figure 3 reveals that all membranes are asymmetric where the pores are smaller on the top of the membrane and became gradually larger at the bottom of the membrane. Figure 3b shows that the membrane has spongy microvoid pores and finger-like macrovoids . This observation is in accordance with the results reported by Lai et al. [10]. This structure was formed during the phase inversion where the PEG and solvent diffused into the non-solvent [25]. The PVDF crystallization occurred very fast during the coagulation process and was especially enhanced by the large temperature difference between the dope (60 °C) and coagulation bath (~22 °C). The dense microvoid structure plays an important role in the separation process of the feed solution. However, compared to the dimension of dextran molecules, some membrane pores are large enough to allow the dextran molecules to pass across the membranes.

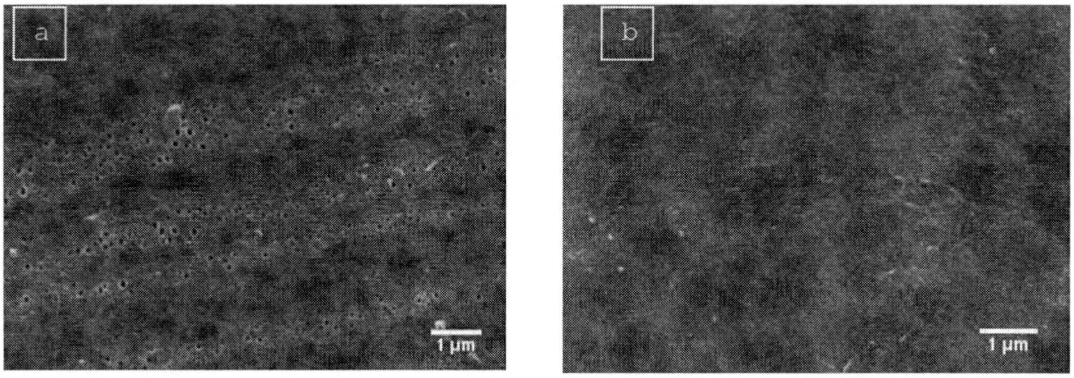

Figure 2. SEM photos of surfaces of membranes with (a) 15% and (b) 17% PVDF concentration

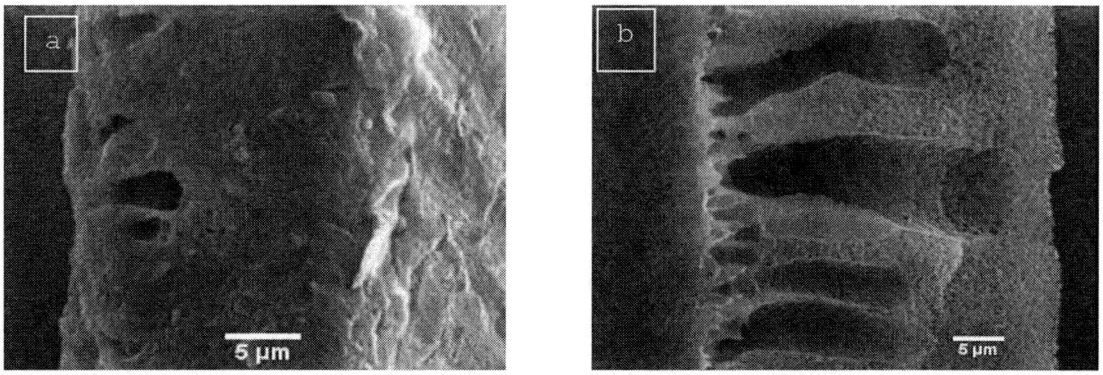

Figure 3. SEM photos of cross-sections of membranes with (a) 15% and (b) 17% PVDF concentration

4. Conclusions

PVDF membranes with various PVDF concentrations have been successfully prepared by phase inversion method. From the experimental data, it can be concluded that the PVDF concentration in the dope affects the surface properties, permeability and permselectivity of the produced membranes. Higher PVDF concentrations results in higher hydrophobicity, mechanical properties and rejection towards T-500 and T-2000 dextrans, but lower water flux.

Acknowledgements

The authors thanks the *LPPM* of Institut Teknologi Bandung for the financial support given through the Research Program, A category, 2016.

References

[1] Pezeshk N, Rana D, Narbaitz R M and Matsuura 2012 *J. Membr. Sci.* **389** 280-6
[2] Huang X, Wang W, Liu Y, Wang H, Zhang Z, Fan W and Li L 2015 *Chem. Eng. J.* **273** 421-9
[3] Velu S, Muruganandam L and Arthanareeswaran G 2011 *int. J. Chem. Analyt. Sci.* **2** 82-6
[4] Sivakumar M, Mohan D R and Rangarajan R 2006 *J. Membr. Sci.* **268** 208-19
[5] Chakrabarty B, Ghoshal A K and Purkait M K 2008 *J. Membr. Sci.* **309** 209-21
[6] Santos W N, Iguchi C Y and Gregorio R 2008 *Polym. Test.* **27** 204-8
[7] He T, Zhou W, Bahi A, Yang H and Ko F 2014 *Chem. Eng. J.* **252** 327-36
[8] Ishigami T, Nakatsuka K, Ohmukai Y, Kamio E, Maruyama T and Matsuyama H 2013 *J. Membr. Sci.* **438** 77-82
[9] Peng Y, Fan H, Dong Y, Song Y and Han H 2012 *Appl. Surf. Sci.* **258** 7872-81
[10] Lai C Y, Groth A, Gray S and Duke M 2015 *Chem. Eng. J.* **267** 73-85
[11] Gijsbertsen-Abrahamse A J, Cornelissen E R and Hofman J A M H 2006 *Desalination* **194** 251–258
[12] Zhao X and Liu C 2016 *Journal of Membrane Science* **515** 29-35
[13] Wongchitphimon S, Wang R, Jiraratananon R, Shi L and Loh C H 2011 *Journal of Membrane Science* **369** 329 - 338
[14] Zhao X and He C 2015 *Appl. Mater. Interfaces.* **7** 17947-53
[15] Kotte M R, Cho M and Diallo M S 2014 *J. Membr. Sci.* **450** 93-102
[16] Kuo C Y, Lin H N, Tsai H a, Wang D M and Lai J Y 2008 *Desalination* **233** 40-7
[17] Shen X, Yin X, Zhao Y and Chen L 2015 *Colloid. Poly. Sci.* **4** 1205–13
[18] Pezeshk N and Narbaitz R M 2012 *Desalination* **287** 247-54
[19] Dong C, Dai Y, Jiang S and He G 2017 *J. Environ. Chem. Eng.* **5** 877-83
[20] Liu D, Li D, Du D, Zhao X, Qin A, Li X and He C 2015 *J. Membr. Sci.* **493** 243-51
[21] Wang P, Ma J, Wang Z, Shi F and Liu Q 2012 *Langmuir* **28** 4776 - 4786
[22] Dong C, He G, Li H, Zhao R, Han Y and Deng Y 2012 *J. Membr. Sci.* **387-388** 40-7
[23] Yan L, Shui Y, Bao C and Xianda S 2006 *J. Membr. Sci.* **276** 162-7
[24] Kim J H and Lee K H 1998 *J. Membr. Sci.* **138** 153-63
[25] Smolders C A, Reuvers A J, Boom R M and Wienk I M 1992 *J. Membr. Sci.* **73** 259-75

State Equation Determination of Cow Dung Biogas

A Marzuki[1], L B Wicaksono

Department of Physics, Sebelas Maret University, Surakarta 57126 Indonesia

[1]amarzuki@mipa.uns.ac.id

Abstract. A state function is a thermodynamic function which relates various macroscopically measurable properties of a system (state variable) describing the state of matter under a given set of physical conditions. A good understanding of a biogas state function plays a very important role in an effort to maximize biogas processes and to help predicting combation performance. This paper presents a step by step process of an experimental study aimed at determining the equation of state of cow dung biogas. The equation was derived from the data obtained from the experimental results of compressibility (κ) and expansivity (β) following the general form of gas state equation $dV= \beta dT + \kappa dP$. In this equation, dV is gas volume variation, dT is temperature variation, and dP is pressure variation. From these results, we formulated a unique state equation from which the biogas critical temperature (T_c) and critical pressure were then determined ($T_c = 266.7\ K$, $P_c =$ 5096647.5 Pa).

1. Introduction

The equation of state of a gas is a mathematical expression describing a state of a gas under a given set of physical conditions. Having this equation, one can formulate a unique relationship between gas state variables such as temperature, volume, pressure, gas concentration, or internal energy [1, 2, 3]. Since this relation can be microscopically associated with a complex interaction of all the particles composing the gas, different gas or different gas mixture will result in different equation of state [4]. In fact, until now no single equation can describe behavior of all gasses. Even, one gas or one gas mixture under different physical condition is described with different equation of state.

Knowing equation of state of a gas or a mixture of gasses plays a very important role in many industrial processes especially on calculating phase equilibrium. By exploiting a gas equation of state, a minimum condition for achieving a phase equilibrium can be obtained [5, 6, 7]. Any thermodynamic processes carried out in such condition can thus save energy. In addition, by exploiting a gas equation of state, one can determine whether any processes can proceed under certain condition.

Biogas processing is one of interesting example that has drawn much attention [8, 9]. Upon combustion, biogas which mainly contains methane produces energy. Unfortunately, biogas cannot be treated in the same way as liquid petroleum gas. So far, biogas is used as a fuel both for cooking and powering motor vehicles in the form of compressed gas. Nowdays, improving biogas processing method as well as on studying biogas mixtures are intesively researched. This paper presents a simple method to derive the equation of state of cow dung biogas from simple experiment to determine gas compressibilty and expansivity. From this equation, the important gas thermodynamic property, critical point, can be obtained.

2. Experiment

Figure 1 is experiment setup used to measure cow dung biogas isobaric expansivity and isothermal compressibility. It mainly contains gas container (1), water bath (2), thermometer (3), syrinx (4), load (5) and Mercury filled U Pipe (6). Gas container is filled with cow dung gas such that for a given mass load, piston was raised to a particular scale.

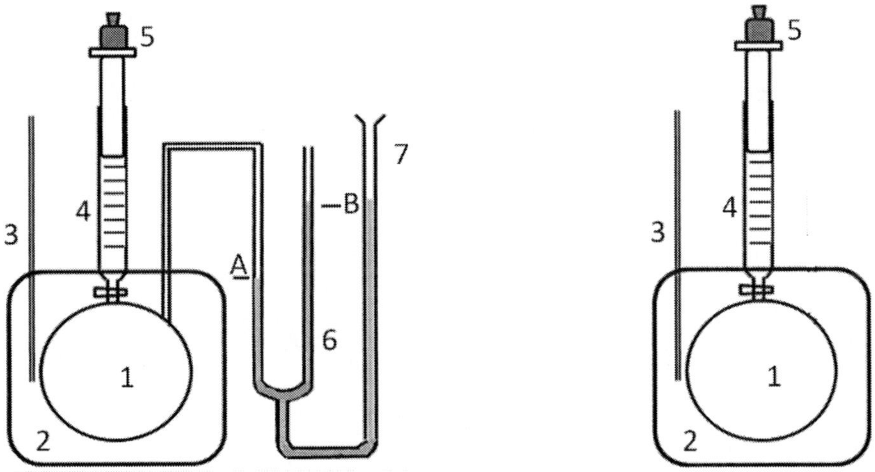

Figure 1. Experimental setup used to measure (a) isobaric expansivity and (b) isothermal compressibility of cow dung biogas. Set up consists of gas container (1), water bath (2), thermometer (3), syrinx (4), load (5) and Mercury containing U Pipe (6).

Figure 1 (a) is experimental set up used to measure isothermal compressibility carried out by following equation (2). Experiment was performed at several constant temperatures. Each carried out as function of mass load. The volume change resulted from a given mass load was read directly from syrinx scale after bringing the mercury level at the left side of the U pipe to point A. At this condition, the magnitude of the gas pressure was calculated from the different height of mercury surface levels (between A point and B point) using equation:

$$P = \rho g h \tag{1}$$

Where ρ, g, and h are mercury density, gravitational acceleration, and different height between points A and B, respectively.

Figure 1 (b) is isobaric expansivity experiment carried out by following equation (3). Constant gas pressure was created by keeping the mass load constant. As gas container was heated up to a constant temperature, the gas volume was change to some degree that can be read directly from syrinx scale.

3. Results and Discussion

3.1. *Isothemal Expansivity*

Isothermal expansivity experiment was carried out by observing the relation between cow dung temperature (T) and its volum change (ΔV) (Figure 2). The way the volume change as function change is best fitted (R^2 = 0.9981) with the exponential function as

$$\Delta V = a + b\frac{exp(kT) - 1}{k} \tag{2}$$

IOP Conf. Series: Earth and Environmental Science 75 (2017) 012028 doi:10.1088/1755-1315/75/1/012028

Where a, b, and k are -0.02375, 4.63047 x 10^{-9}, and 0.9981, respectively. Remembering that the isobaric expansivity (β) is defined as

$$\beta = \frac{1}{V}\left(\frac{\partial V}{\partial T}\right)_P \tag{3}$$

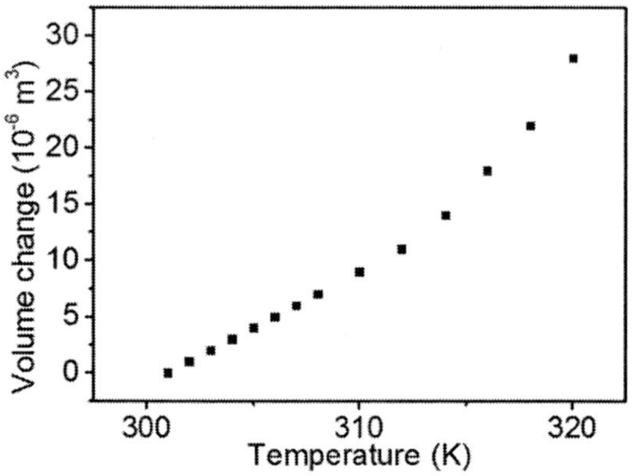

Figure 2. Volume change of cow dung gas as function of temperature measured at a constant pressure.

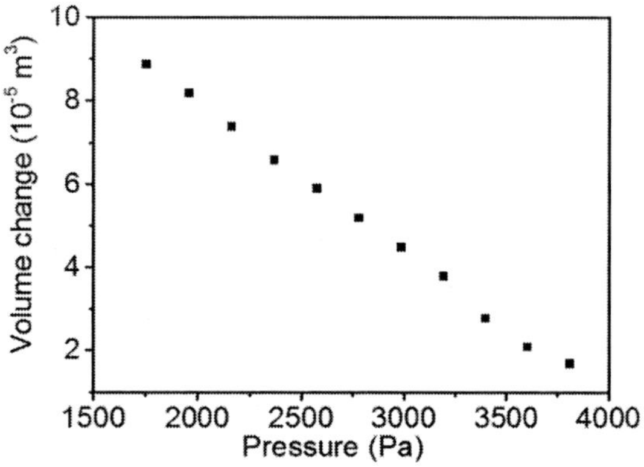

Figure 3. Volume change of cow dung gas as function of pressure measured at a constant temperature.

From equation (2), one can obtain

$$\beta = \frac{1}{\boldsymbol{a + \frac{b}{k}\,(e^{(kT)}) - \frac{b}{k}}} \cdot \frac{d}{dT}\,(a - \frac{b}{k} + \frac{b}{k}\,e^{(kT)})$$

or

$$\beta = \frac{(e^{(kT)})}{(\frac{a}{b} - \frac{1}{k} + \frac{1}{k}e^{(kT)})} \tag{4}$$

3.2. *Isothermal compressibility*

Isothermal compressibility is obtained in a similar way to that of finding isobaric expansivity. From figure 3, the curve is best fitted (R^2 = 0.99759) with the equation:

$$\Delta V = \boldsymbol{c\,exp}\left(\frac{\boldsymbol{-P}}{\boldsymbol{d}}\right) + \boldsymbol{fP + e} \tag{5}$$

Where c = 0.52, d = 18088.80, e = -0.36 and f = -1.09 x 10^{-5}. Inserting equation (5) to the isothermal compressibility equation difined as

$$\kappa = -\frac{1}{V}\left(\frac{\partial V}{\partial P}\right)_T \tag{6}$$

We found

$$\kappa = \frac{c.e^{\left(-\frac{P}{d}\right)} - fd}{dc\, e^{\left(-\frac{P}{d}\right)} + ed + fdP} \tag{7}$$

3.3. State Equation and Critical Point

State equation of any gas can be obtained from thermodynamic relation

$$dV = \left(\frac{\partial V}{\partial T}\right)_P dT + \left(\frac{\partial V}{\partial P}\right)_T dP \tag{8}$$

Inserting equation (3) and (6) into equation (7) we found

$$dV = \beta V\, dT - k\, V\, dp$$

$$\frac{dV}{V} = \beta\, dT - K\, dp \tag{9}$$

Equation (9) is general state of equation. If isobaric expansivity (β) and isothermal compressibility (κ) are experinetally known, a state equation relating P, V, and T characterizing a gaseous state of a material can be derived. Bearing this in mind, the state equation of cow dung biogas can be derived by inserting equation (4) and (7) into equation (9):

$$\frac{dV}{V} = \frac{b.exp(kt)}{\left(a - \frac{b}{k} + \frac{b}{k} exp(kT)\right)} dT - \frac{c.exp\left(-\frac{P}{d}\right) - fd}{dc\, exp\left(-\frac{P}{d}\right) + ed + fdP} dP \tag{10}$$

we found

$$\ln V = \ln\left[dc\ \boldsymbol{exp}\left(-\frac{P}{d}\right) + ed + fdP\right] + \ln\left[\left(\frac{a}{b} - \frac{1}{k} + \frac{1}{k} exp(kT)\right)\right] \tag{11}$$

or

$$V = \left[dc\ \boldsymbol{exp}\left(-\frac{P}{d}\right) + ed + fdP\right]\left[\left(\frac{a}{b} - \frac{1}{k} + \frac{1}{k} exp(kT)\right)\right] \tag{12}$$

Using state equation expressed by equation (12), one can then calculate the very important parameters usually used in a process dealling with phase transformation from gas phase to liquidus phase. They are critical temperature (T_c), critical pressure (P_c), and critical volume (V_c). In pressure volume diagram (P-V diagram), critical point is isothermal line deflection point. At any temperatures above this isothermal line, a gas cannot be liquified no matter how much pressure is applied. Since citical point is a deflection point, this point can be calculated from the mathematical conditions

$$\frac{\partial P}{\partial V} = 0 \text{ and } \frac{\partial^2 P}{\partial^2 V} = 0 \tag{13}$$

Inserting equation (12) into equation (13) and utilizing the reciprocal equation

$$\left(\frac{\partial V}{\partial P}\right)_T = \frac{1}{(\partial P/\partial V)_T} \tag{14}$$

We found $T = T_c = 266.7$ K and $P = P_c = 5096647.5$ Pa. With this result, we can see that cow dung biogas cannot be liquified for the purpose of gas packaging in the same way as liquid petroleum gas (LPG) at room temperature. A new mixture of biogas with critical point higher than room temperature has to be found. For this purpose, critical point of each mixture can be calculated from gas equation of state derived from simple compressibility and expansivity experiment.

4. Conclusion

Cow dung biogas has been thermodynamically studied in order to derive its state equation. The equation was derived based on experimental results of isobaric expansivity and isothermal compressibility. We found that the equation of state of cow dung biogas is $V = \left[dc\ \boldsymbol{exp}\left(-\frac{P}{d}\right) + ed + fdP\right]\left[\left(\frac{a}{b} - \frac{1}{k} + \frac{1}{k}exp(kT)\right)\right]$. From this equation we calculated the critical temperature and pressure: $T_c = 266.7$ K, $P_c = 5096647.5$ Pa. From this experiment, it can be seen that cow dung biogas cannot be liquified for the purpose of gas packaging in the same way as liquid petroleum gas (LPG) at room temperature.

References

[1] Pérez-Polo M F, Pérez-Molina M, Varó E F and Chica J G 2016 *Chem. Eng. Sci,* **155** pp. 482-503.
[2] Amooey A A 2014 *Chin. J. Chem. Eng.* **22** 6 pp. 628-633.
[3] Jarrahian A and Heidaryan E 2014 *Fuel* **134** pp. 333-342.
[4] Guerrero-Zárate D, Estrada-Baltazar A and Iglesias-Silva G A 2017 *Fluid Phase Equilib.* **437** pp 69-82.
[5] Huang T, Li C, Jia W and Peng Y, Fluid *Phase* Equilib. **427** pp. 35-45.
[6] Petropoulou E, Pappa G D and Voutsas E 2017 *Fluid Phase Equilib.* **433** pp 135-148.
[7] Zhao W L, Zhong D L and Yang C 2016 *Fluid Phase Equilib.* **427** pp. 82-89.
[8] Yousef A M, Eldrainy Y A, El-Maghlany W M and Attia A 2017 *J. Nat. Gas Sci. Eng.* **45** pp. 812-824.
[9] Ugarte P, Durán P, Lasobras J, Soler J, Menendez M and Herguido J 2017 *Int. J. Hydrogen Energy* **42** pp. 13589-13597.